Photographic Atlas of the Mid-Atlantic Ridge Rift Valley

Robert D. Ballard
James G. Moore

Photographic Atlas

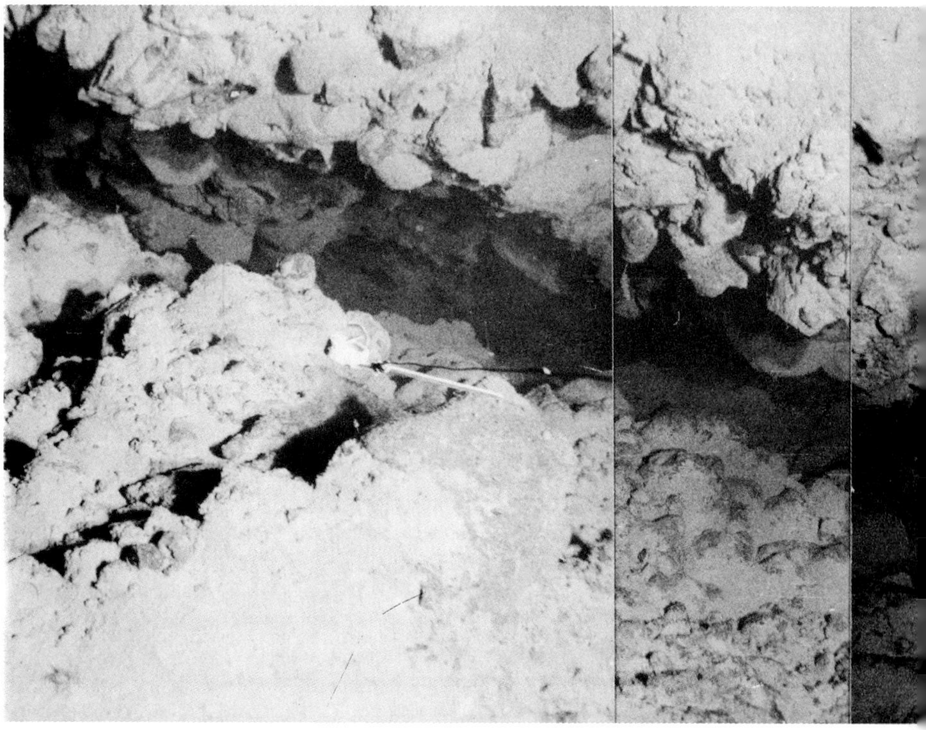

of the Mid-Atlantic Ridge Rift Valley

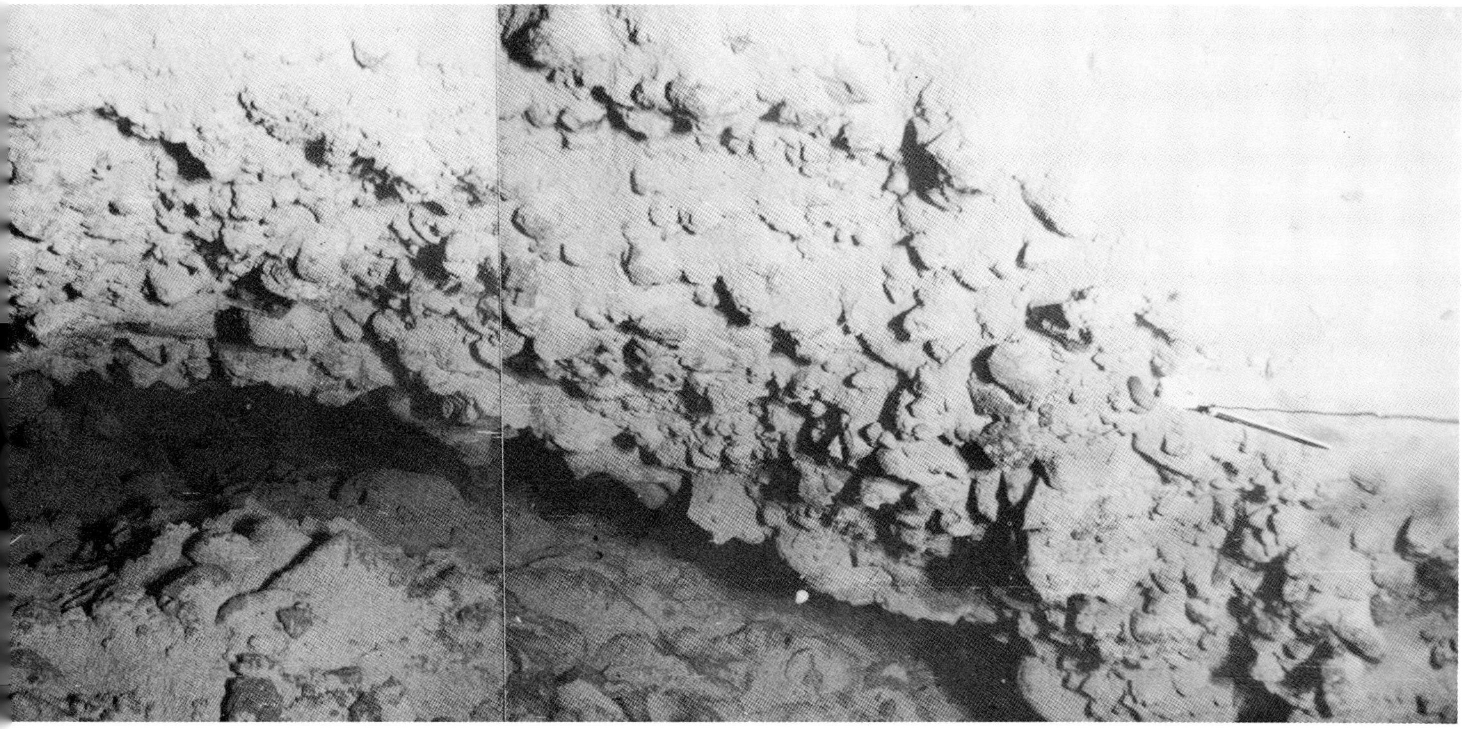

Springer-Verlag New York Heidelberg Berlin

Robert D. Ballard
Woods Hole Oceanographic Institution
Woods Hole, Massachusetts 02543

James G. Moore
U. S. Geological Survey
Menlo Park, California 94025

Library of Congress Cataloging in Publication Data

Ballard, Robert D.
Photographic atlas of the Mid-Atlantic Ridge Rift Valley

Bibliography: p.
1. Mid-Atlantic Ridge Rift Valley—Maps, Pictorial.
2. Submarine topography—Atlantic Ocean—Maps, Pictorial.
3. Submarine Valleys. 4. Photography, Submarine.
I. Moore, James Gregory, 1930—joint author. II. Title.
G2807.M5B3 1977 912'. 19' 636 77-25187
ISBN 0-387-90247-3

Printed in the United States of America.
9 8 7 6 5 4 3 2 1

ISBN 0-387-90247-3 Springer-Verlag New York

ISBN 3-540-90247-3 Springer-Verlag Berlin Heidelberg

Dedicated to the crews, scientists, engineers, and pilots who first demonstrated in Project FAMOUS that diving craft can safely perform detailed scientific studies of the deep ocean floor.

Foreword

The oceans are so large and our knowledge of them so limited that we sometimes think of the sea floor as a vast uniform wasteland. But modern oceanographic research is revealing that each part of the sea floor has its own characteristic features and is as distinct as the geologic and topographic provinces on land. The deep ocean floor holds the same fascination for us as unexplored mountain tops, jungles, or the surface of the moon and planets. And, since the oceans cover more than 70 percent of the earth's surface, the ocean floor holds the key to much of the history and evolution of our earth.

There are very few places in the depths of the ocean that have been explored and studied directly by man, and to which man can relate in the way he relates to the earth on land. The Mid-Atlantic Ridge rift valley, studied during Project FAMOUS, is one of these places. This location was chosen because it is typical of the zone in the central Atlantic where volcanic activity apparently creates all of the sea floor of that ocean. Consequently the region is of great importance to earth scientists. Here participating scientists systematically investigated large and small geologic features to learn their shape, properties, and origin. They used the most advanced techniques available from aircraft, surface ships, and submersibles which operated in water depths of nearly 3 kilometers. As part of these studies an extensive photographic record of bottom features was obtained. Some of these photos are reproduced in this book to give the reader a view of the sea floor and to help him understand the natural processes at work.

Robert Ballard is a geologist with extensive experience in submersible diving and James Moore, a volcanologist who has made many scuba dives in active volcanic terrain. Both were members of the American scientific diving team of the project. They, and their colleagues, have published nearly 100 papers on the scientific results of the program. However, these papers have reproduced only a few of the large volume of photographs that are now in the archives. We believe that the public at large as well as our scientific colleagues will benefit from the selected photographs in this volume.

Woods Hole
November, 1977

J. R. Heirtzler

Acknowledgments

The photographs in this atlas were obtained through the combined cooperation of the American and French scientists, engineers, technicians, crews, and pilots associated with Project FAMOUS. The authors wish to express their sincere appreciation to this international group who worked long hours at sea during the four year period of the project.

We wish to particularly thank Drs. Jean-Louis Cheminee and Gilbert Bellaiche of the Centre National de la Recherche Scientifique and Walter Brundage of the U.S. Naval Research Laboratory for their assistance in the preparation and review of this atlas. FAMOUS was funded in the United States by the National Science Foundation and the National Oceanic and Atmospheric Administration with support from the Office of Naval Research, and the Naval Oceanographic Office. Chief scientist for the American phase of Project FAMOUS was Dr. James R. Heirtzler, Woods Hole Oceanographic Institution. The French program, operating under the government agency Centre National pour L'Exploitation des Oceans, CNEXO, was directed by Mr. Claude Riffaud and Dr. Xavier Le Pichon.

We also wish to thank Susan Christiansen and Irene Fisher for the sketches they made of various pillow and tectonic forms contained within this atlas.

Robert D. Ballard
James G. Moore

Contents

Introduction

Approximately 200 million years ago, the continental nuclei of Europe, Africa, and the Americas began to separate as a result of dynamic processes taking place within the earth (Figure 1). As the land masses moved apart, a large crack or rift zone developed in the Earth's rigid outer layer, the lithosphere. In response to separation of the lithosphere along this zone of rifting, plastic material from the earth's interior rose toward the surface filling the void. Unlike the lighter granitic rock which forms the core of most continents, the rising material from the upper mantle crystallized to form basaltic and related denser rock. Since the lithosphere of the earth is floating on denser, more plastic deeper layers, this newly created basaltic rock initially equilibrated at a level lower than the lighter granitic continents nearby. The result was a depression, the Atlantic Ocean, situated between the separating continental masses. This depression was filled by the invading waters of the Tethys Sea (now Mediterranean) and Pacific Ocean which existed at that time.

Although the young Atlantic floor was lower than the adjacent continents, it still retained a great deal of its original heat, particularly adjacent to its site of formation at the spreading rift. As the new lithosphere was rafted outward from the rift, it slowly cooled and contracted. The youngest or most recently crystallized crust was swollen with heat, producing a welt in the form of a mountain range situated above the site of upwelling and rifting. As separation continued, the older crust was transported laterally farther and farther away from its original site of formation (Figure 2). In time it cooled and contracted, sinking to greater depths to form the deep ocean floor located between the continents and the younger mountain range in the center of the ocean.

The process of separation in the Atlantic has been occurring at a relatively

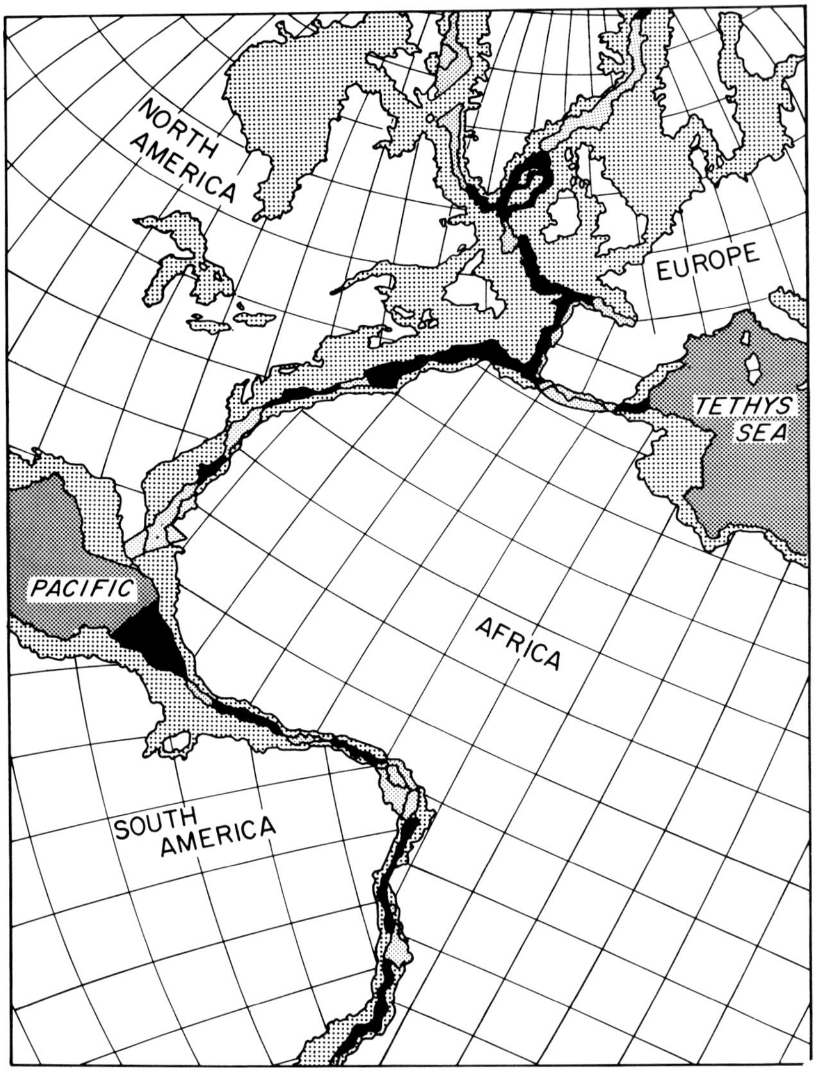

1 A geometrical fit of the continental nuclei around the Atlantic Ocean which is thought to represent the rough outline of the supercontinent of Pangea which existed approximately 200 million years ago. This reconstruction was made by Bullard et al., 1965 using a least squares numerical method of fitting. The best fit was found to occur at the 500 fathom contour on the continental margins of the bordering land masses (lined pattern). The black areas correspond to voids in the fit whereas the small dotted pattern is where overlap occurs. It is along this boundary of separation that the Mid-Atlantic Ridge first came into existence.

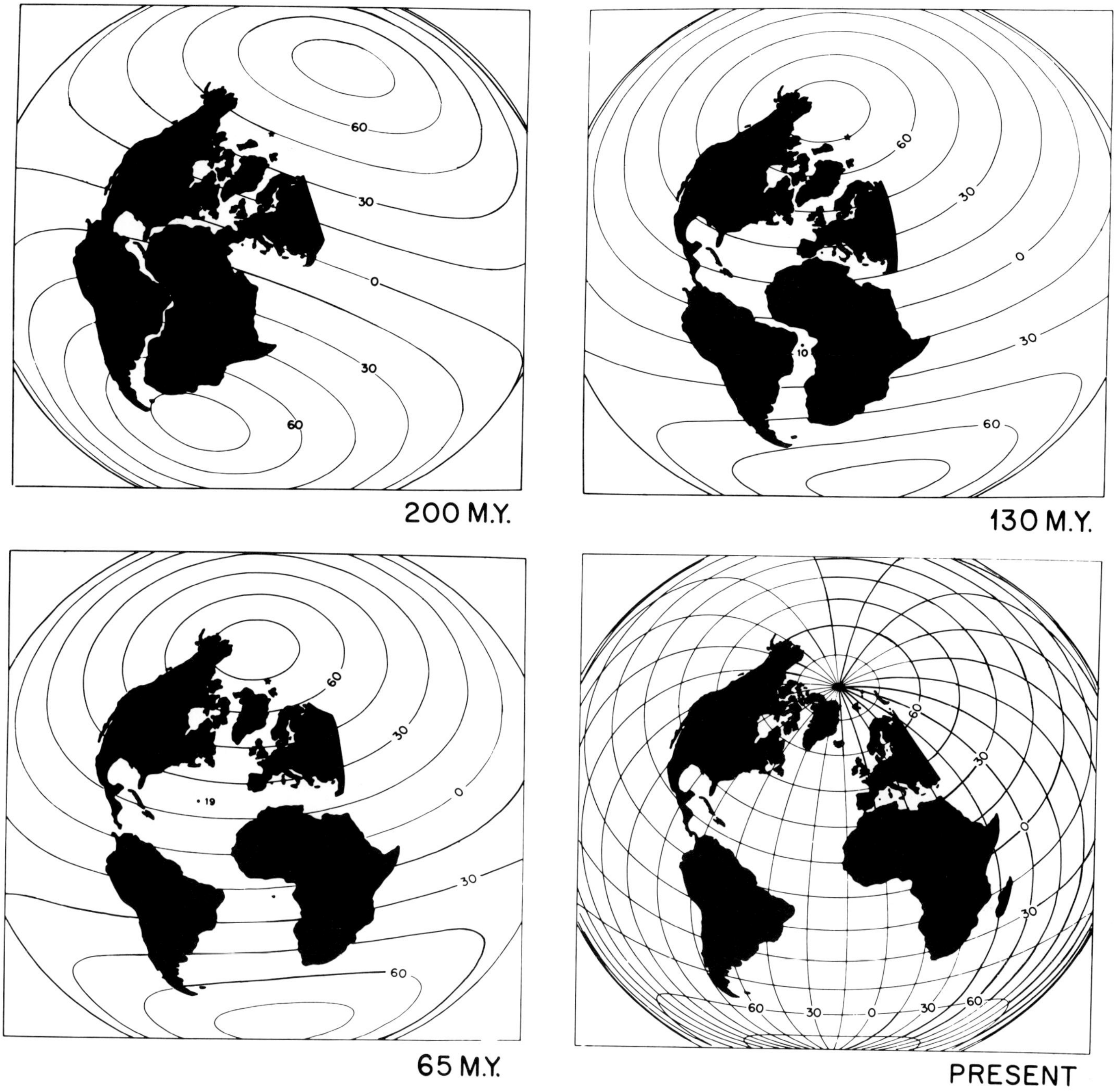

2 Using Bullard et al. (1965) poles of rotation, Phillips and Forsyth (1972) generated a series of reconstructions showing the evolution of the Atlantic Ocean as we know it today. Their work suggests that the central Atlantic, Caribbean, and Gulf of Mexico began to form 200 million years ago while the South Atlantic first opened as late as 150 million years ago.

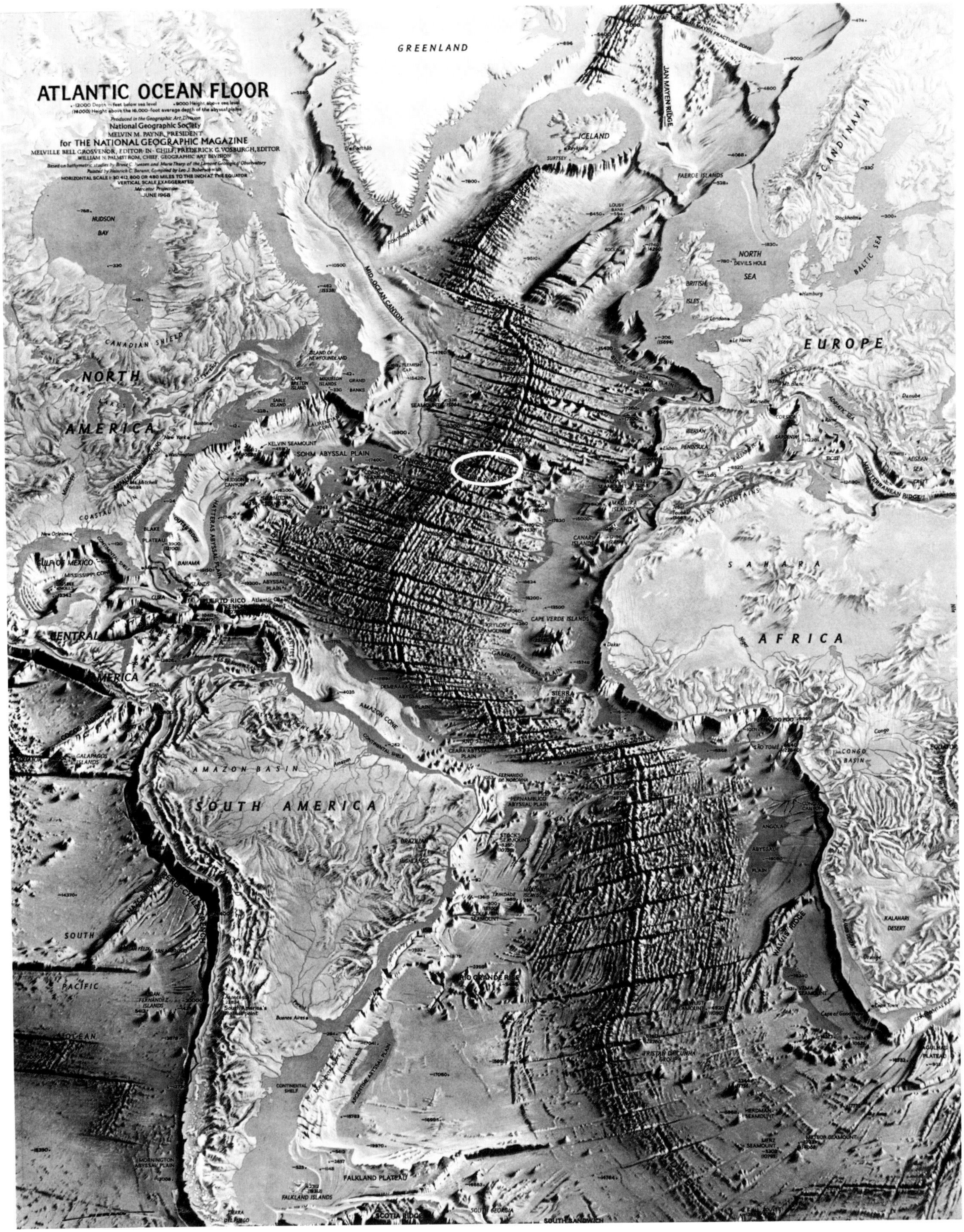
ATLANTIC OCEAN FLOOR
-12000 Depth in feet below sea level
9000 Height above sea level
(14000) Height above the 16,000-foot average depth of the abyssal plains
Produced in the Geographic Art Division
National Geographic Society
MELVIN M. PAYNE, PRESIDENT
for THE NATIONAL GEOGRAPHIC MAGAZINE
MELVILLE BELL GROSVENOR, EDITOR-IN-CHIEF; FREDERICK G. VOSBURGH, EDITOR
WILLIAM N. PALMSTROM, CHIEF, GEOGRAPHIC ART DIVISION
HORIZONTAL SCALE 1:30,412,800 OR 480 MILES TO THE INCH AT THE EQUATOR
VERTICAL SCALE EXAGGERATED
JUNE 1968
GREENLAND
ICELAND
JAN MAYEN RIDGE
SCANDINAVIA
FAEROE ISLANDS
HUDSON
BAY
MID-OCEAN CANYON
NORTH SEA
BRITISH ISLES
BALTIC SEA
CANADIAN SHIELD
EUROPE
NORTH
AMERICA
GRAND BANKS
SOHM ABYSSAL PLAIN
KELVIN SEAMOUNT
IBERIAN PENINSULA
AEGEAN SEA
BLAKE PLATEAU
HATTERAS ABYSSAL PLAIN
CANARY ISLANDS
GULF OF MEXICO
SAHARA
NARES ABYSSAL PLAIN
CAPE VERDE ISLANDS
AFRICA
CENTRAL
AMERICA
GAMBIA ABYSSAL PLAIN
SIERRA LEONE RISE
AMAZON CONE
CONGO BASIN
GALAPAGOS ISLANDS
AMAZON BASIN
SOUTH AMERICA
PERNAMBUCO ABYSSAL PLAIN
ANGOLA
BRAZILIAN HIGHLANDS
KALAHARI DESERT
SOUTH
PACIFIC
OCEAN
RIO GRANDE RISE
WALVIS RIDGE
Buenos Aires
TRISTAN DA CUNHA
HERDMAN SEAMOUNT
MERZ SEAMOUNT
FALKLAND PLATEAU
FALKLAND ISLANDS
SCOTIA RIDGE
SOUTH GEORGIA
SOUTH SANDWICH

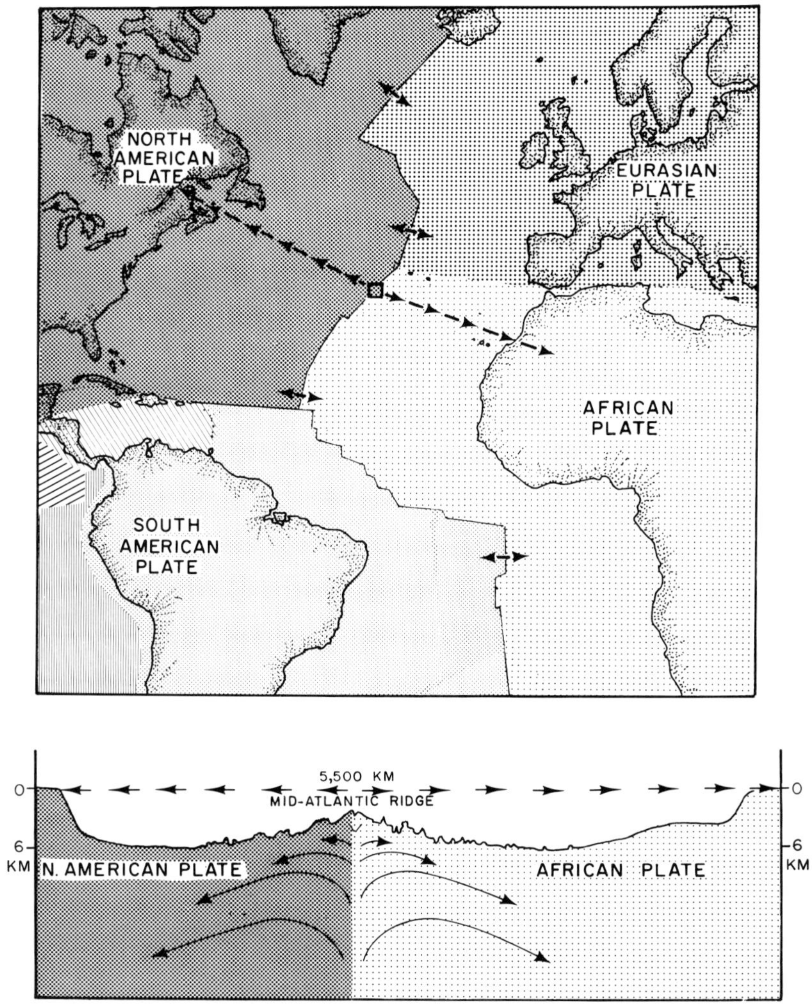

4 The top of the figure shows the crustal plates surrounding the Mid-Atlantic Ridge. The arrows denote the principal direction of relative plate motion and the small square is where the Project FAMOUS study was conducted. The lower portion of the figure is a generalized profile extending from the continental shelf of North America across the Mid-Atlantic Ridge rift valley to the continental shelf of northwest Africa.

3 Physiographic map of the Atlantic Ocean sea floor and surrounding land masses. Published and copyrighted by the National Geographic Society and based upon bathymetric studies by Bruce C. Heezen and Marie Tharp. Running down the center of the Atlantic is the Mid-Atlantic Ridge. The boundaries of the major plates meeting along this ridge line are marked by the rift valley extending along the crest of the mountain range. The small ellipse shows the region of the Mid-Atlantic Ridge where Project FAMOUS was conducted.

continuous rate for the intervening 180 million years. The central mountain range of freshly forming oceanic lithosphere of today is the Mid-Atlantic Ridge (Figure 3). Parallel to the crest of this mountain range and at its center is a deep rift valley, caused by the pulling away of two large lithospheric plates: the American plate which is moving west and the African plate which is moving east (Figure 4). The floor of this rift valley is mantled with very young basaltic lava flows which are fed from the upwelling hot mantle material.

As each episode of rifting or sea-floor spreading occurs, a portion of the plastic upper mantle material rises and because of reduced confining pressure undergoes partial melting. This partial melt rises towards the surface where it is probably stored temporarily in a shallow chamber. Finally a portion of the magma eventually flows out onto the sea floor as basaltic lava flows at a temperature of approximately 1,250°C. When this hot lava comes into contact with the cold seawater of the ocean floor, it is quickly cooled into a tangled mass of lobate flow tongues about the size and shape of pillows. Such "pillow lava" forms thicker master flow units which are commonly stacked on one another and build small, steep-sided volcanoes.

Immediately after formation this volcanic terrain is subjected to contractional cooling and the local tensional forces of sea-floor spreading. The newly formed crust is cracked and faulted as it is transported laterally to make room for the next intrusive and volcanic cycle.

Such faulted volcanic terrain is the bedrock which underlies most of the world oceans. It represents the most common form of volcanic products on earth, but until recently was unavailable for systematic study. In 1972 a comprehensive marine field program called Project FAMOUS (French-American Mid-Ocean Undersea Study) was initiated to investigate the volcanic and tectonic processes taking place in the inner rift valley of the

Mid-Atlantic Ridge (Heirtzler and LePichon, 1974). During the course of this four-year program, a unique series of bottom photographs was obtained. These photographs portray the original volcanic features as well as their subsequent modification by tectonic processes. Many of the photographs were obtained firsthand by scientists on the rift valley floor in small deep diving submersibles. The remainder were taken using surface-lowered remote camera units. Documentation and interpretation of observations, photographs, and samples have been published in a variety of journals. The purpose of this photographic atlas, however, is to present a visual aspect of the terrain and geological features occurring in the rift valley. Photographs of the various features are grouped together so as to emphasize the similarity or modification of important forms and structures. Since the principal emphasis of this atlas is on the oceanic crust exposed within the rift valley, little mention is given to the abundant marine life observed or areas where the sediment cover was thick enough to bury the bedrock surface.

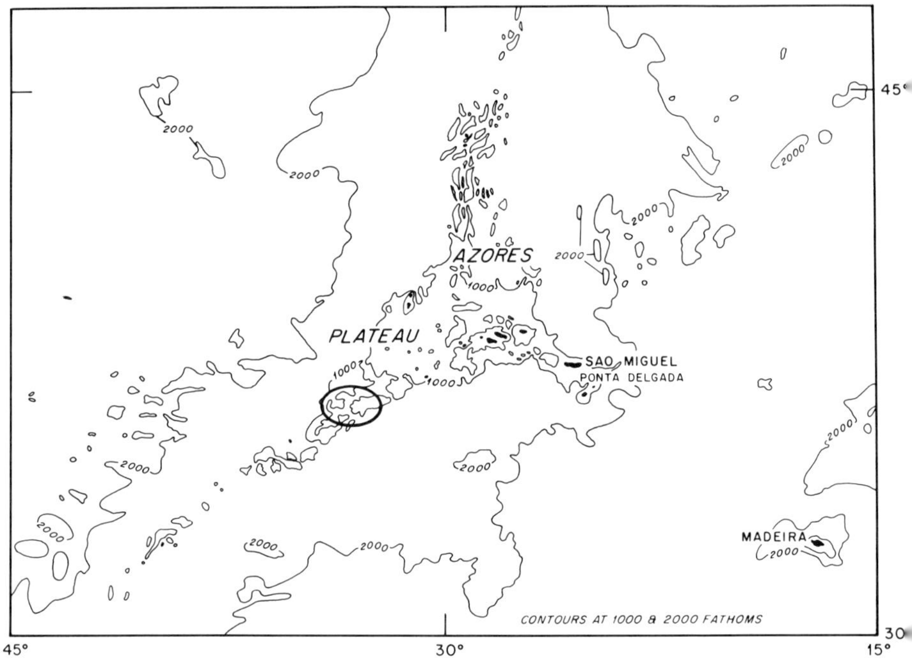

5 The Project FAMOUS study area (ellipse) is situated on the southwestern portion of the Azores Plateau near the island of San Miguel which served as the base of operations for the program.

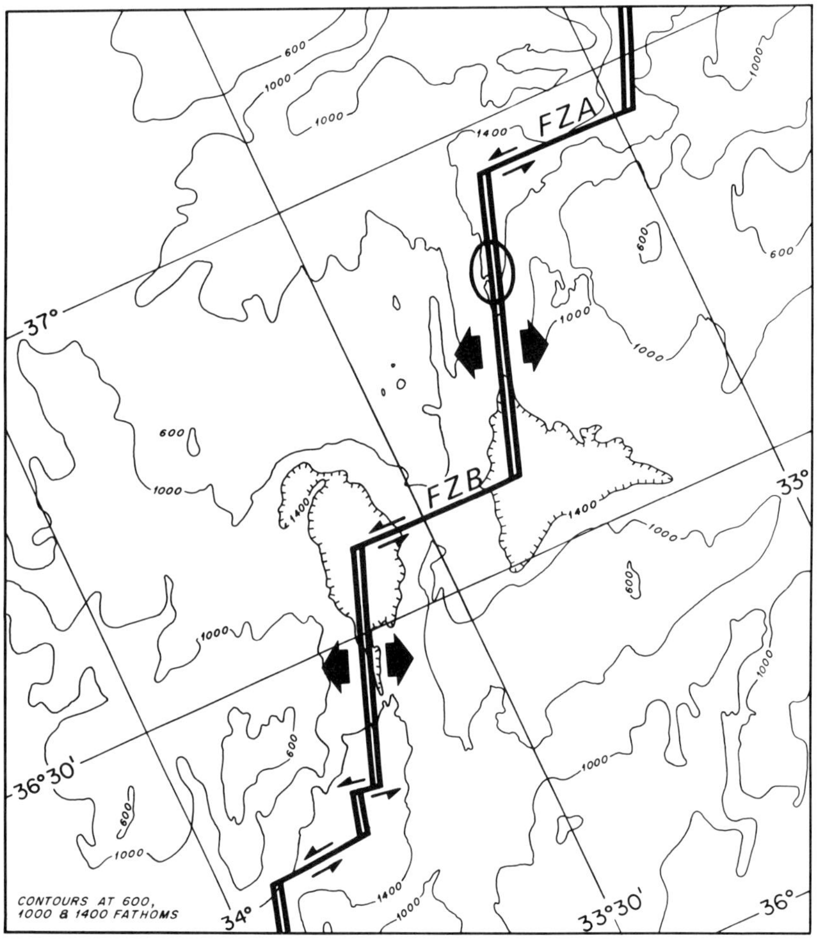

6 A further enlargement of the study area. The bathymetric data were obtained from a detailed series of maps compiled by the Naval Oceanographic Office (Phillips and Fleming, in press). At this scale it can be seen that the Mid-Atlantic Ridge consists of a series of short rift valley segments connected by intersecting fracture zones. The rift valley segments mark the boundary of plate separation (large arrows) while the fracture zones represent areas of faulting and crustal translation (small arrows). The ellipse is the area where the majority of the ALVIN dives were conducted.

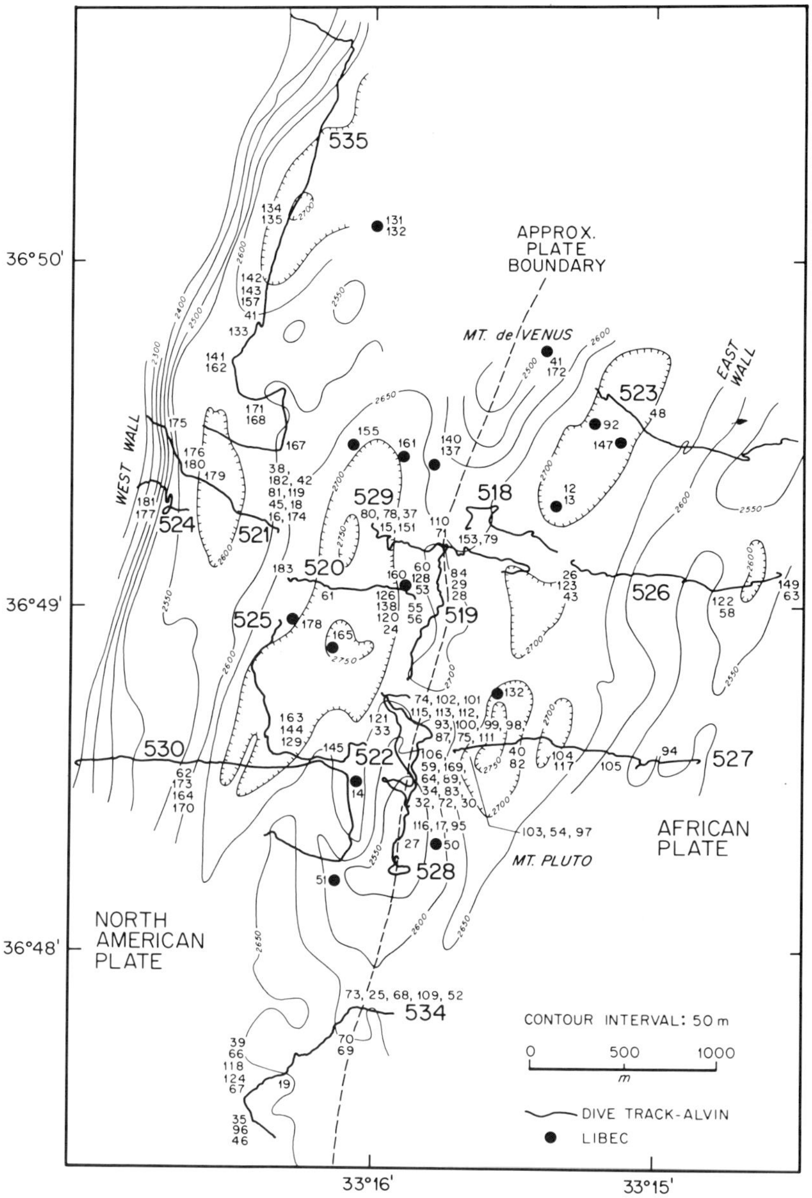

7 The small-scale topography of the rift valley floor where the photographs in this atlas were taken. Mt. Venus, Mt. Pluto, and Mt. Uranus are volcanic highs extending down the axis of the valley and representing the area of recent volcanic activity. This central zone is flanked by the remainder of the floor which consists of older volcanic units. The valley itself is flanked by the east and west walls which consist of inward facing fault scarps. Large numbers are ALVIN dive numbers and small numbers indicate the approximate location of photo figures in this atlas.

Geologic Setting

The Project FAMOUS study area is located along a portion of the Mid-Atlantic Ridge rift valley which intersects the southeastern margin of the Azores plateau (Figures 4 and 5). The area is situated some 720 km southwest of Ponta Delgada on the Azores island of San Miguel.

The rift valley does not extend continuously along the crest of the ridge, but instead is divided into a series of segments 20–45 km in length trending more nearly north-south (N29° E). The individual valley segments are offset by small horizontal-slip faults called transform faults which intersect the rift valley at approximately right angles (Figure 6). In general the inner valley can be subdivided into three geologic provinces (Figure 7). The first is a central zone of recent volcanic activity which outlines the present boundary between the two separating crustal plates (Figure 8). The most common features in this zone are young volcanic ridges which exhibit only recent faulting, faulting of small displacement which has not greatly modified the primary volcanic features. The second province includes the valley floor to either side of the central volcanoes. In this region the volcanic terrain is modified by small to intermediate scale fault scarps. Pelagic sediment a few meters thick has accumulated in the low-lying areas. The third province contains the east and west inner walls of the rift valley where vertical uplift is taking place on a major scale. It is on these fault scarps that the deeper internal layers of the oceanic crust are exposed.

Photographic Techniques

The photographs shown in this atlas were taken using two techniques. The majority were obtained from the three-man submersible ALVIN (Figure 9). This

U.S. Navy submersible is operated by the Woods Hole Oceanographic Institution and has a diving limit of 3,650 meters (12,000 feet). The photographs were taken with both exterior cameras mounted on the submersible and internal cameras held by the scientists inside the pressure sphere and focused through its viewports. The external cameras are mounted on the brow of ALVIN and look forward. Pictures taken with these cameras can be distinguished from other photographs by the presence of the sample basket, a compass, and frequently a portion of a mechanical arm in the lower part of the frame (Figure 9). Scale in these photographs can be estimated from the fact that the water samplers on the sample basket have a length of 25

8 An artistic rendering of a portion of the rift valley shown in Figure 7. Using the detailed topography of the area an artist for the National Geographic Society rendered the sea floor relief based upon discussions with the various geologists who traversed the valley in ALVIN. Mt. Venus is shown on the right and Mt. Pluto on the left. The scarp in the background is the first major vertical step in the west wall. Copyright National Geographic Society.

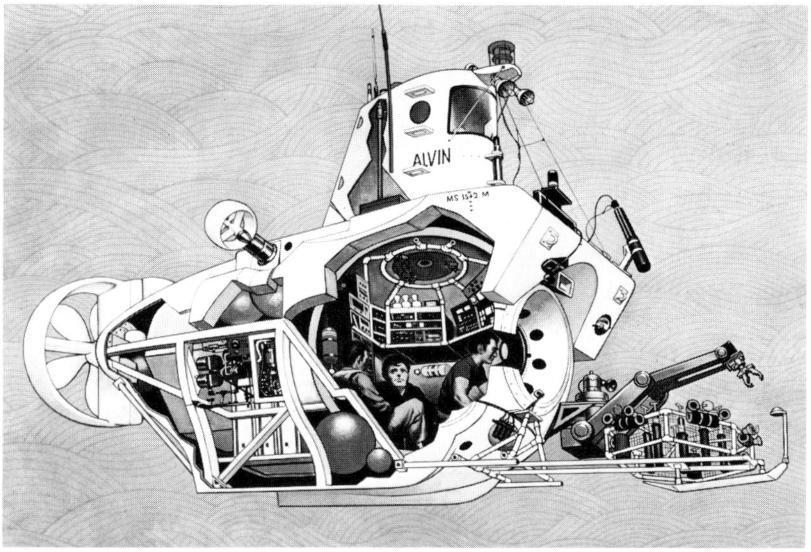

9 The major internal and external components of the submersible ALVIN which include a titanium alloy pressure sphere, mechanical arm, sample tray with tools, external television and still cameras with lights, and horizontal and vertical propulsion units. © 1975 National Geographic Society.

cm and a diameter of 10.5 cm. The external 35 mm cameras have a 50 mm focal length lens and are synchronized to a 200-watt strobe. Standard settings were shutter speed 1/25 second, and f-stop, 5.6. When the submersible is moving at 1/4 knot, 2 to 3 meters above the bottom, a shutter repetition rate of 10 seconds provides a series of photographs having 10–20 percent overlap. Some of the photographs in the atlas are crude mosaics constructed from such overlapping photographs. The mosaics have not been rectified and therefore contain some distortion.

The photographs taken by the scientists within the pressure sphere were made with NIKON 35 mm cameras having 24 and 43–83 mm zoom lens. The cameras were synchronized to an external 500 watt-s strobe. The highest quality pictures were taken at an f-stop of 5.6 within a distance of 1–3 meters from the viewport. The principal films used in all cameras were Plus-X and 2496 black and white shot at their normal ASA settings.

The majority of the remaining photographs shown in the atlas were taken by a surface-towed camera unit developed by the U.S. Navy called the LIBEC system (*LI*ght *BE*hind the *C*amera) (Figure 10). This 70 mm camera has a seawater focal length of 40.4 mm. The normal distance off the bottom was 12 meters, and each photograph covers an area approximately 22 meters across. Suspended beneath the camera and within its field of view is a compass which was used to orient the features observed relative to the axis of the valley.

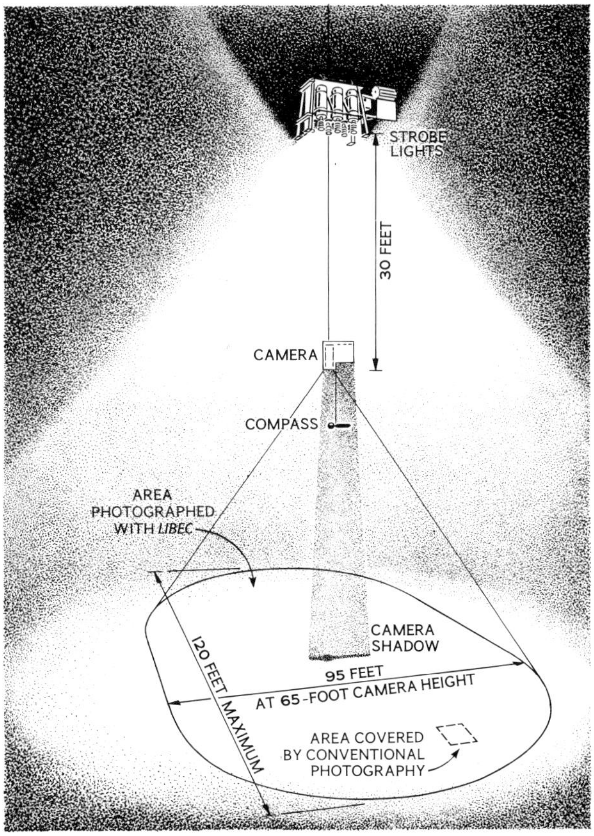

10 Diagram of the LIght BEhind the Camera (LIBEC) system showing the electronic flash lamps with camera suspended beneath. The area of the bottom obtained in a photograph is shown as well as the average photographic coverage of conventional camera units (after Brundage and Cherkis, 1975). © 1975 National Geographic Society

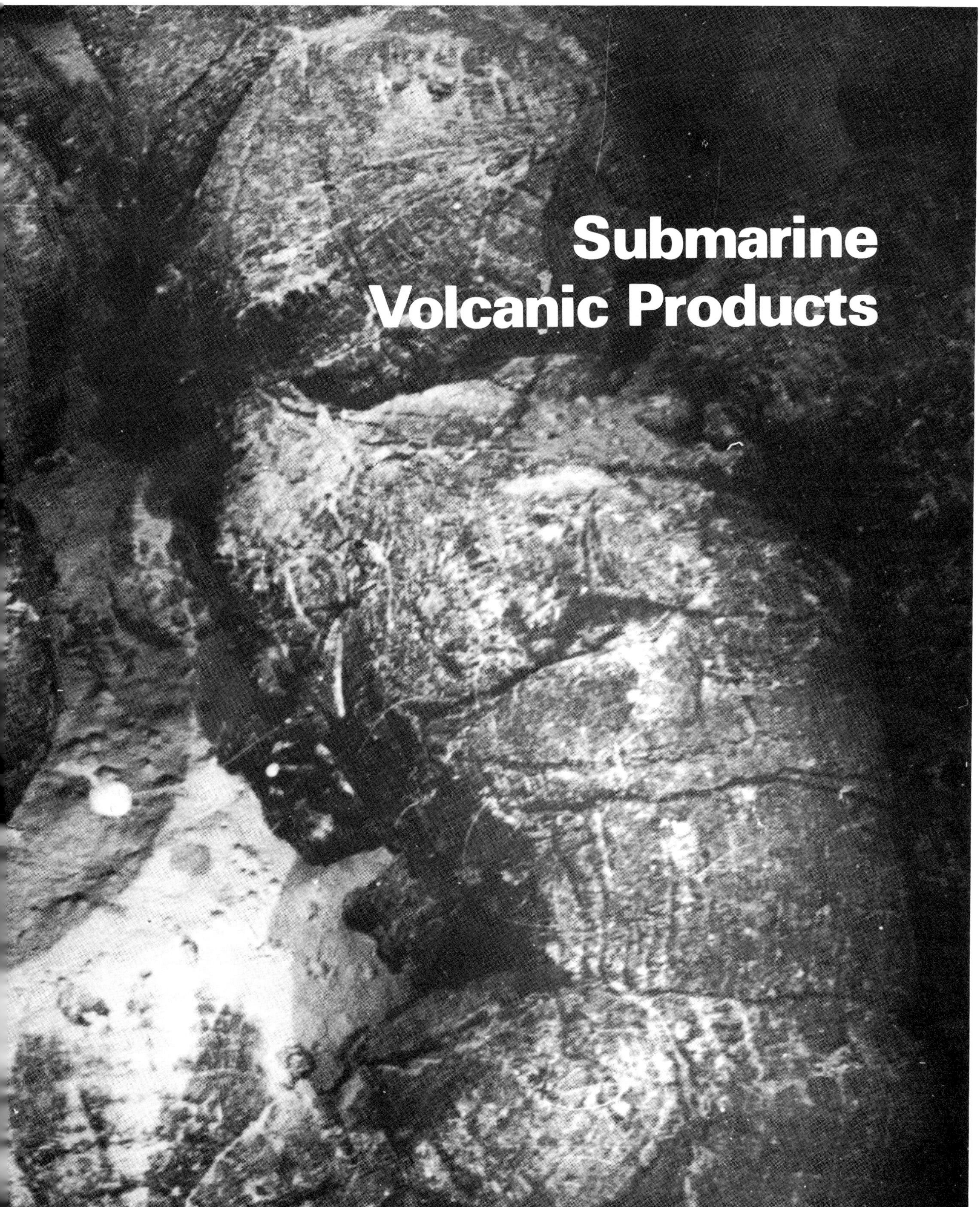
Submarine
Volcanic Products

11 Sketch of pillowed cone at site of volcanic vent. Note radially draped elongate pillows and lack of summit depression.

The great bulk of new ocean floor created at oceanic spreading ridges, such as the Mid-Atlantic Ridge in the FAMOUS area, is composed of pillowed basaltic lava flows. Such lava flows have been previously studied at the spreading ridges by examination of dredge hauls and photographs taken from equipment lowered from surface ships. Outward from the spreading axis, where younger sediments cover the volcanic terrain, the deep sea drilling program of the GLOMAR CHALLENGER has verified that the dominant layer underlying the sediment is composed of pillow lava. This characteristic type of lava also is dominant in the submarine part of

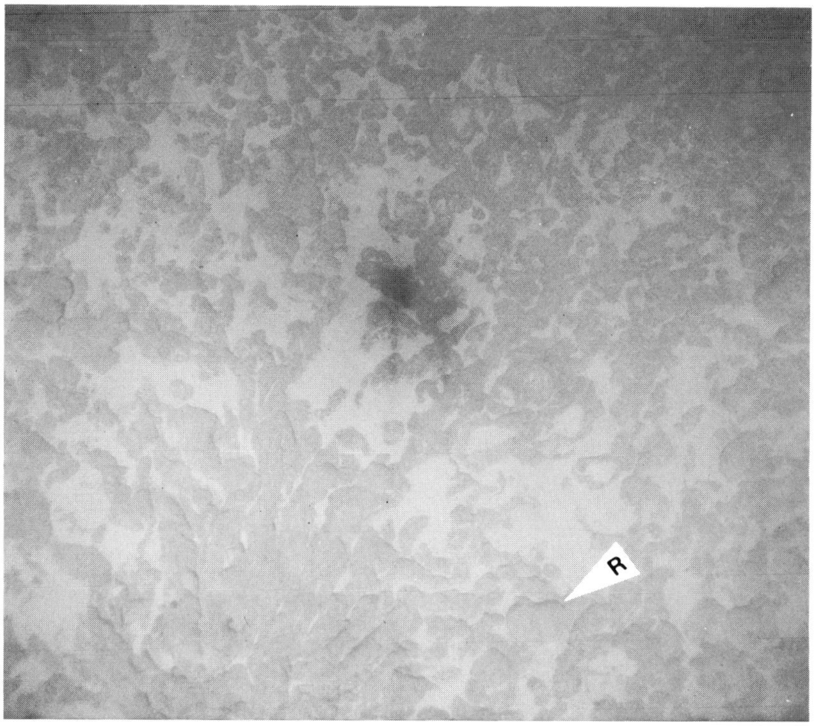

mid-ocean volcanoes such as Hawaii, and is important in the submerged part of the island arc volcanoes. Pillowed lava flows erupted beneath water likely comprise the most voluminous form of volcanic rock on earth.

Submersible observations have shown that the inner floor of the FAMOUS rift valley is composed of a great variety of young pillowed basaltic flows (Heirtzler and Bryan, 1975; Bellaiche and Cheminee, 1976), whereas similar volcanic rocks occur outward from the main fault scarps bounding the inner valley, but are obviously older, more weathered, and are largely covered by young sediments. Hence the locus of active extrusion of lava is within the inner rift valley, which is 2–3 km wide.

This lava is apparently erupted through tensional cracks of the inner rift valley floor which parallel the axis of the valley. The cracks are formed by fracture of the thin, brittle roof of an underlying magma chamber in response to the divergence of the two lithospheric plates which meet along the rift valley axis. As these plates move apart by the process of sea-floor spreading, the cracks widen and deepen until they tap magma at depth which erupts to the surface. Although the lava is relatively fluid, it flows only a short distance from the vent because of rapid chilling by seawater and tends to pile up above the vent.

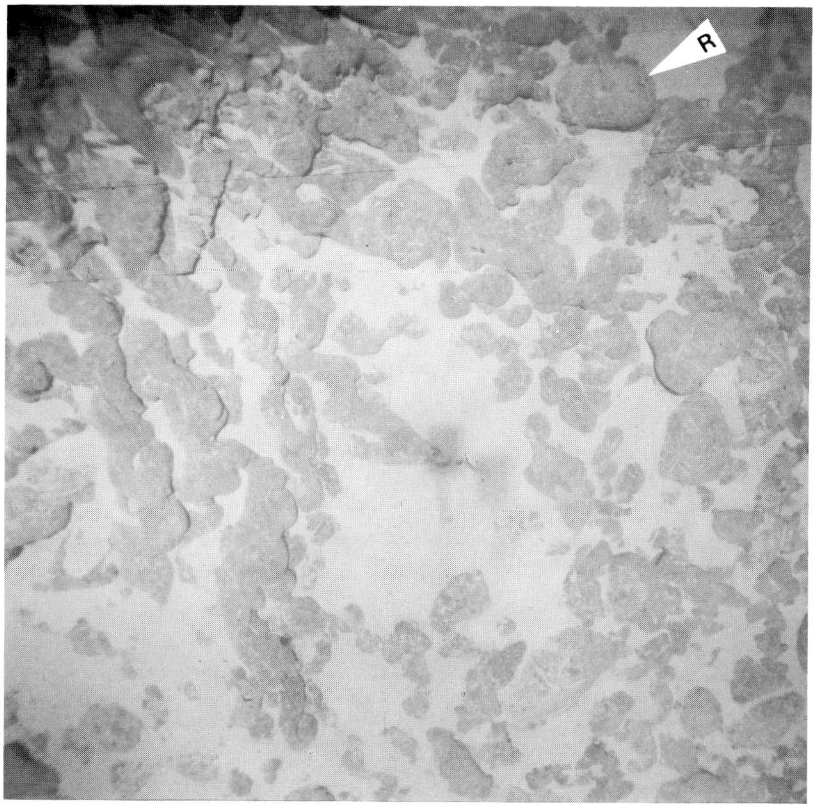

12 Vertical LIBEC photograph showing pillowed cone marking volcanic vent at bottom margin. Note radial arrangement of elongate pillows fanning out from summit of cone. Large pillow (R) is also shown in following photograph. Narrow dimension of photograph is roughly 21 m, and radius of cone, 10 m.

13 Vertical LIBEC photograph showing flank of pillowed cone of preceding photograph and same large pillow (R). Narrow dimension of photograph is roughly 15 m and longest elongate pillow lobes extend about 12 m from summit.

Continued activity causes lava to repeatedly break through this pile of thick pillowed flows until a ridge several hundred meters high is built in favorable areas, particularly in the center of the inner rift valley. Eventually, this pile seals off the vent and inhibits further extrusion, until further extension and cracking induces a new volcanic phase. New lithosphere is created at the plate juncture by the combined processes of volcanism at the surface and intrusion of molten and plastic material at depth.

Most of the volcanic activity is confined to a narrow zone only a few hundred meters wide in the center of the inner rift valley. Here it has built a series of volcanic hills or ridges rising 100 to 240 m above the flanking areas of the valley floor. Two of these volcanic ridges, Mt. Venus and Mt. Pluto, occur in the FAMOUS areas. However, young volcanic rocks are also associated with other vent areas marginal to the central volcanic ridges. At these sites small volcanic hills only a few tens of meters high have been built above erupting vents.

The FAMOUS project has provided the most complete look yet obtained of basaltic volcanic forms on the ocean floor. Not only were prolonged observations made and thousands of photographs taken in situ from

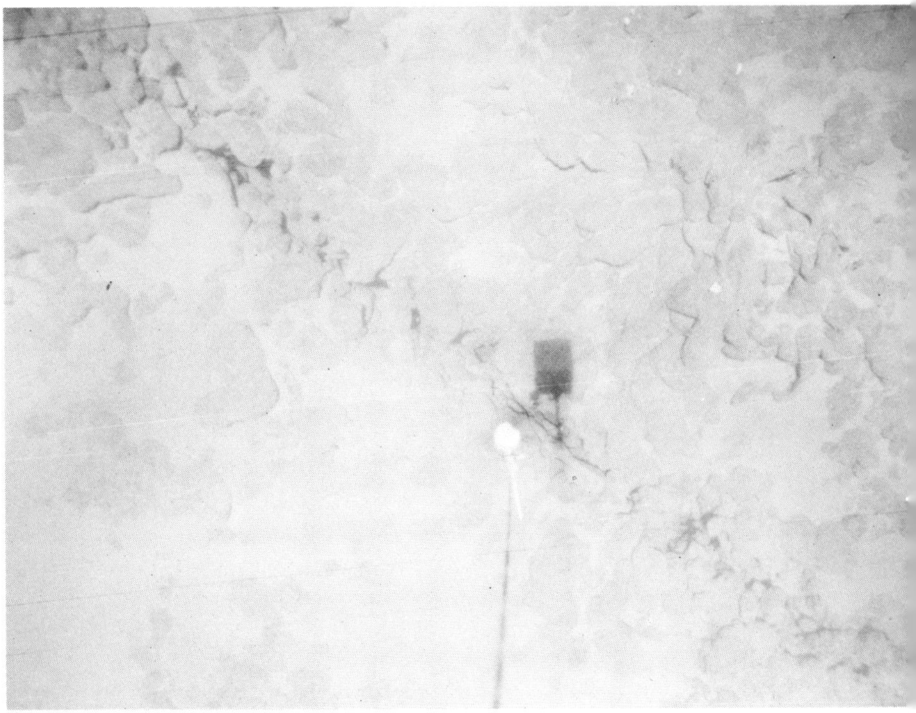

14 Vertical LIBEC photograph with diagonal fissure across the center. Small pillowed cone occurs in upper right with radially arranged elongate pillows. Narrow dimension of photograph is roughly 17 m, and radius of cone, 6 m.

15 Complex pillow on summit of pillowed cone. Animal life such as the siliceous sponge (about 15 cm tall) and the wire-like coral commonly grow in such exposed places. Pillow is surmounted by small trapdoor pillow.

16 Mosaic of elongate pillows on flank and base of pillowed cone.

submersibles, but many photographs were also taken by surface ships including large-area LIBEC photos. The larger volcanic features were investigated by submersible profiling, mosaicing of large area photographs, and analysis of the narrow-beam Navy bathymetry. Finally dredge hauls and submersible sampling provided a look at the small details of well-located individual specimens.

Almost all of the lava forms that were observed are pillowed lava flows. No clear-cut aa or blocky lava flows were seen, and no evidence of explosive eruptions producing pyroclastic rocks was observed. The rich variety of pillow forms provides new insight into the mechanism of eruption, flow, and cooling of basaltic lava on the ocean floor. Consequently, in this section we attempt to illustrate and describe some of the distinctive products of submarine volcanism and introduce a nomenclature of these common, but little understood, volcanic forms.

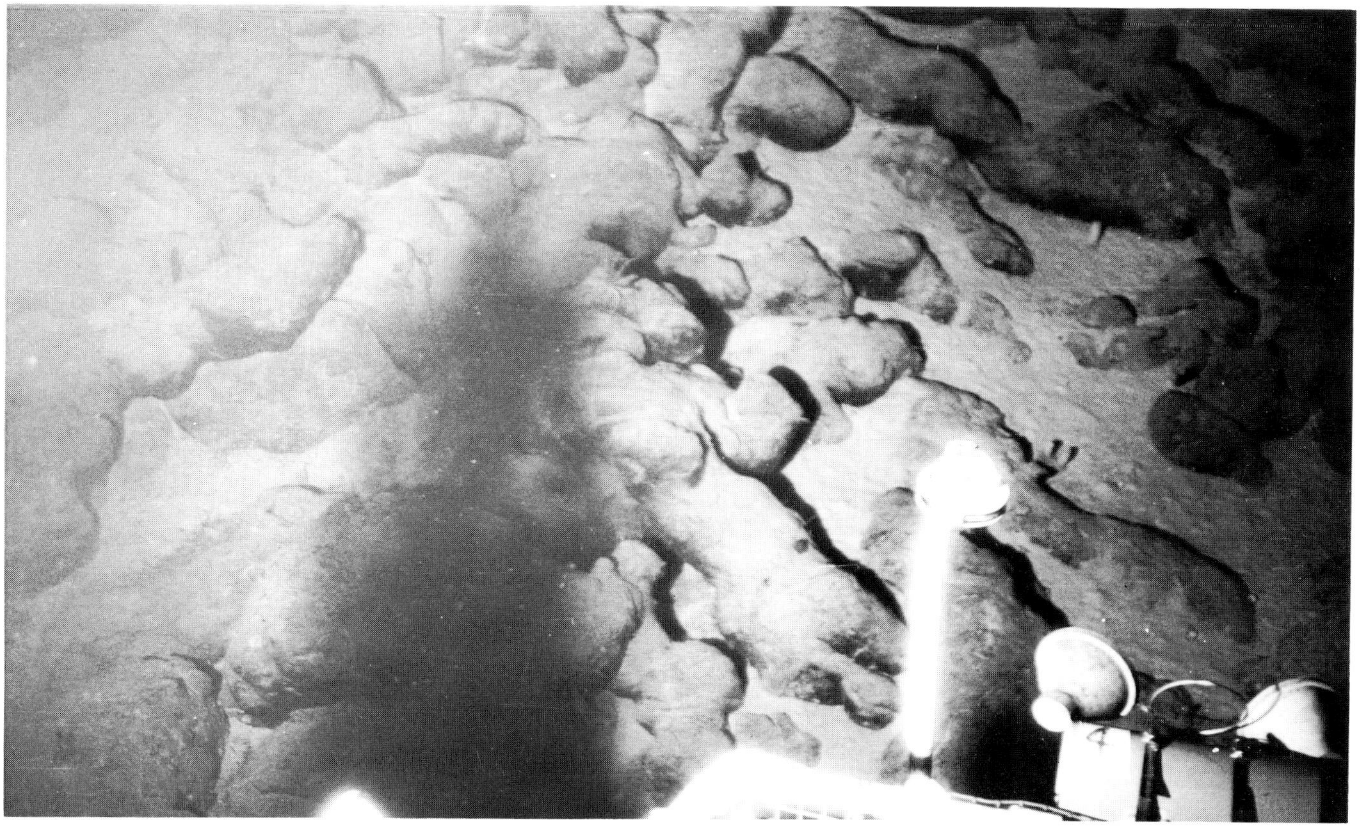

Volcanic Vents

17 Summit of pillowed cone on summit ridge of Mt. Pluto.

18 Lower slope of pillowed cone. Trapdoor pillow in foreground.

19 Slope of pillowed cone.

Photo reconnaissance from the ocean surface revealed that much of the freshest lava occurs in the central axial part of the inner rift valley on top of a series of volcanic ridges including Mt. Venus and Mt. Pluto. These photographs showed no actual open volcanic vents or craters. During the course of subsequent diving operations, the crests of Mt. Pluto and Mt. Venus were carefully investigated and no open vents were seen. The sites of volcanic extrusion are marked by piles of lava resembling a haystack. These *pillowed cones* are 5–20 meters high and are composed entirely of outward-dipping, radially-arranged elongate pillow lobes (Figure 11). They are generally surmounted by a single bulbous pillow or an irregular pillow-like mass of lava (Figure 15). The pillowed cones are variable in slope and diameter. The slope is commonly 45 degrees near the top and becomes nearly horizontal near

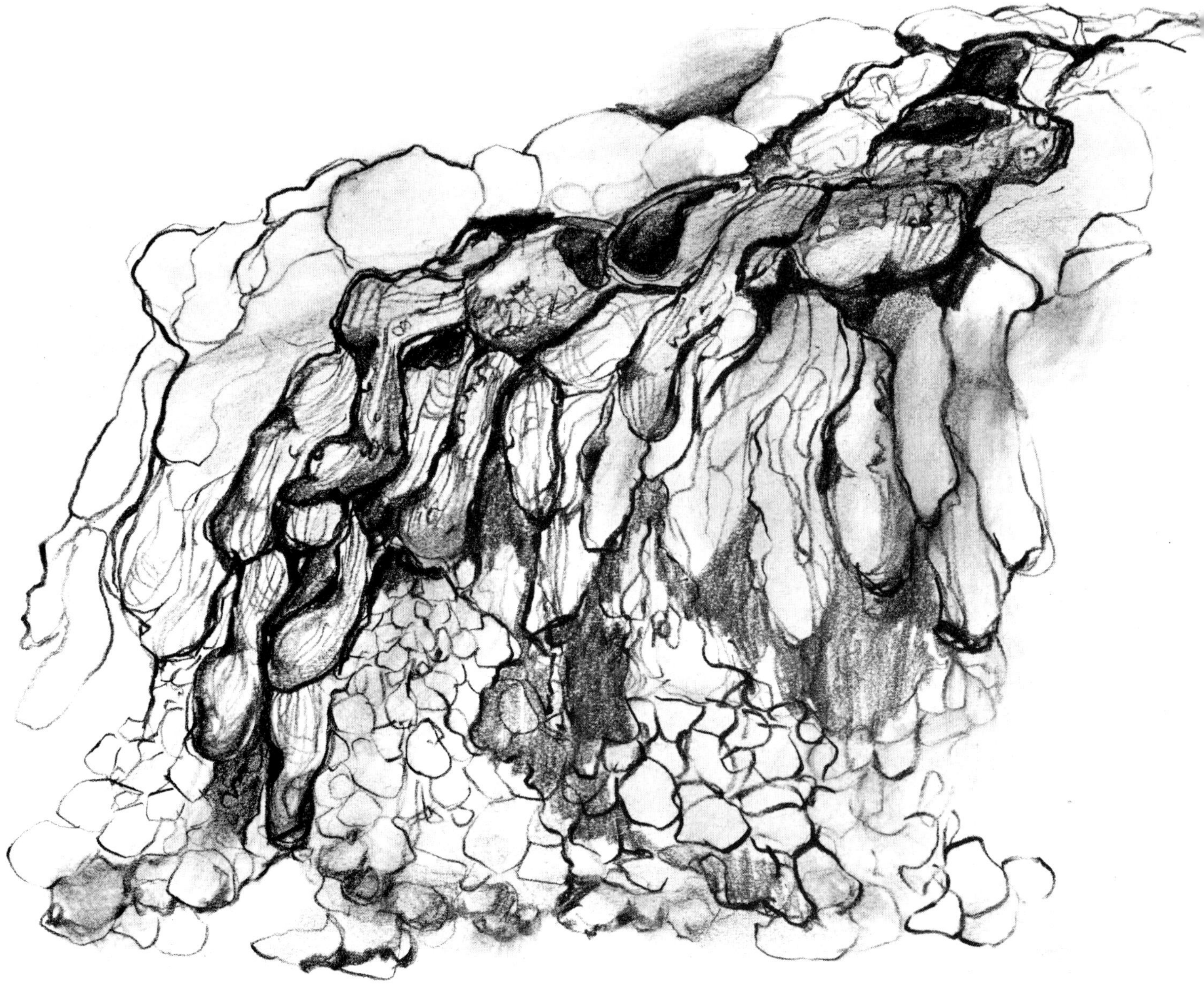

20 Sketch of pillowed flow showing larger bulbous pillows on gentle top surface and smaller diameter foreset-bedded elongate pillows on flow front. Collapsed hollow bulbous pillows on top mark location of underlying master feeding pillow tube. Flow-front rubble composed of broken pillows is formed by slumping of steep flow front.

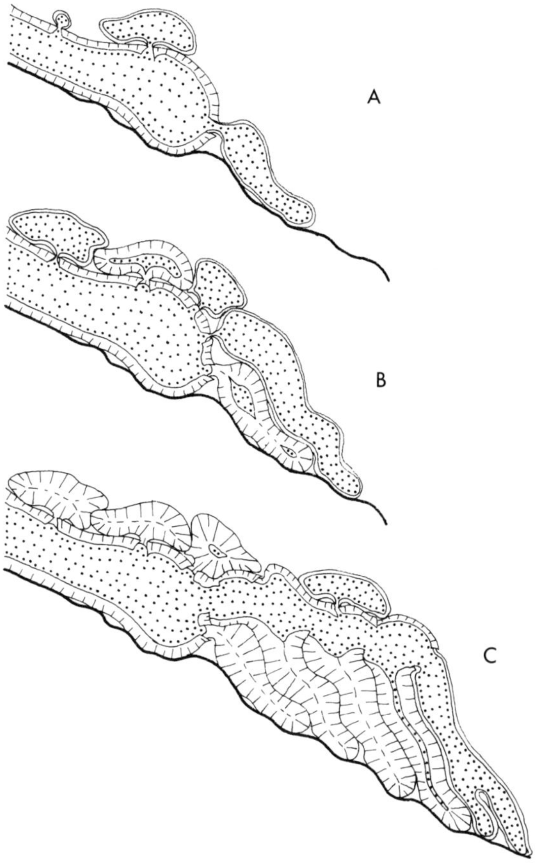

21 Development of foreset-bedded elongate pillows in a pillowed lava flow moving down an irregular slope with a steady supply of new lava. Where the initial flow lobe encounters an obstruction on the sea floor (A), the lobe inflates and ruptures feeding a pillow which drapes the slope. This pillow will cool and stagnate when it reaches a critical size and will effectively dam the advance of the feeding flow lobe, which will then rupture above it (B) and feed a second elongate pillow which drapes the first one. Likewise, bulbous pillows grow from the top of the original flow lobe and expand more equidimensionally because they rest on a more horizontal surface. In this way the flow front is maintained at a steep angle, and a feeding channel is propagated near the top of the advancing pillowed flow (C). The flow is composed of three elements: (1) a top layer of bulbous pillows, (2) a master feeding tube inclined gently downhill from the vent, and (3) a thick basal layer of foreset-bedded elongate pillows plunging at 30–45°. In addition the flow may commonly override flow-front rubble shed from the flow front (not shown). Dotted pattern indicates molten lava.

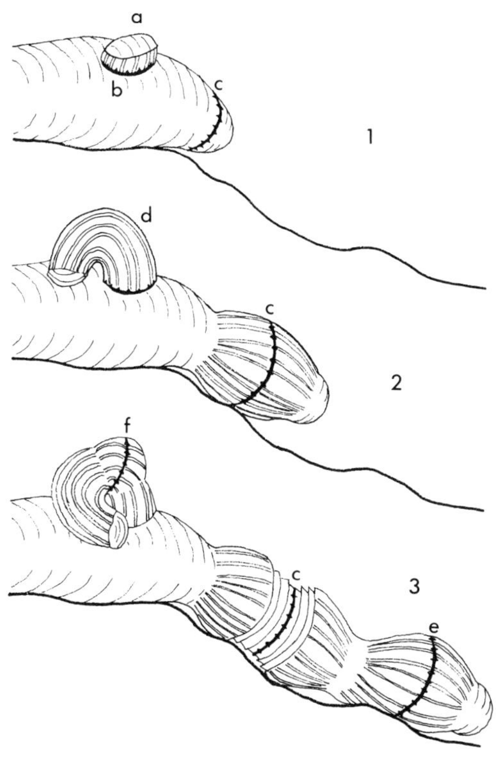

22 Sketch showing growth of interconnected pillows and development of features of pillow crust. Heavy hachured line is incandescent lava welling up into spreading crack; double lines are corrugations. 1. Pillow lobe fed from larger interconnected system at left inflates and fractures at ring crack (b) producing trapdoor pillow (a); end of pillow cracks at (c). 2. Continued growth of trapdoor pillow produces toothpaste pillow (d) which hangs downward because it is largely plastic; spreading on crack (c) pushes original pillow cap downslope developing new corrugated crust on both sides of crack (c). 3. New spreading crack (f) opens on toothpaste pillow developing knob on side. Spreading of crack (c) slows down causing development of faulted, constricted surface, rather than corrugated convex surface. When spreading ceases on crack (c), a new crack (e) develops near end of pillow and rapid spreading forms a new corrugated knob near end of pillow. Continued jumping of spreading crack downslope near end of pillow permits propagation of pillow system downslope producing a complex pinch-and-swell shape of the resulting elongate pillow.

the base where the diameter is from 10 to more than 50 meters (Figures 12–19). Cones also occur on the summits of the volcanic ridges where they have grown on top of larger pillowed flows with steep flow fronts. The summits of the volcanic ridges are composed of a series of pillowed cones arranged in line. The rather constant elevation of such cones suggests that they were fed from interconnected fissures linked to a common magma chamber with a fixed hydraulic pressure.

Further study of the large area LIBEC photographs revealed several pillowed cones in areas not investigated by ALVIN. The radial pattern of these cones is clear in the photographs. Some of the cones may not mark actual volcanic vents, but may have been fed from a master tube in a pillowed flow.

23 Bulbous pillows.

Major Flow Units

Virtually all of the constructional volcanic features of the rift valley floor are composed of thick, stubby, basaltic lava flows composed of numerous individual, rounded flow lobes or pillows (Figure 20). These compound lava flows are termed *pillowed flows.* The pillowed cones are a form of radially arranged pillowed flows at the site of a volcanic vent.

The pillowed flows are generally less than a few hundred meters in length, though they are commonly wider than their length if fed from a fissure vent. They are bounded in the down flow direction by a steep flow front, commonly lobate in plan, with a slope of 30–45 degrees or more and a height of 5 to 20 meters. Slumping of this steep flow front produces piles of rubble at the base (Figure 20). The upper surface of the pillowed flows is markedly hummocky but much gentler in overall slope and averages about 5 degrees in slope from source to flow front. Mt. Pluto, a

24 Field of bulbous pillows on flow top.

25 Large high-standing bulbous pillow.

26 Bulbous pillows partly buried by sediment.

volcanic ridge 200 meters high, is composed of a series of pillowed flows atop one another. The resultant volcanic pile has a distinctly stepped profile. The scarps 5 to 20 meters high are produced by the flow fronts, and the more gentle terraces are underlain by the flow tops.

The individual flow lobes or cooling units are connected to larger masses within the compound pillowed flow (Figure 21). Hence every rounded flow lobe visible on the face of the flow is presumably the end of a long pipe which joins other flow lobes of ever increasing size and eventually connects with the feeding fissure at the source of the flow. These cylinder-like flow lobes are very irregular in size and shape; they curve and change direction abruptly, branch and divide, pinch and swell in diameter, and are decorated with smaller knobs and bulbous masses. Despite this diversity, however, they maintain two distinctive features: a roughly rounded cross-section and a roughly uniform size such that many lobes are 10–100 cm in diameter. The term *pillow* is applied to all or part of these flow lobes which look like a pile of pillows when seen in section.

Because of the tendency for a flow lobe to digitate and produce elongate pillows on steep slopes, a submarine flow can advance by splitting into numerous pillow tongues which drape the steep front (Figure 21). In this manner, foreset-bedded pillows are produced (Fuller, 1931). Figure 21 is a schematic diagram showing the development of foreset-bedded elongate pillows on a submarine lava flow moving down an irregular slope with a steady supply of new lava. Where the flow encounters an obstruction on the sea floor, the front of the flow ruptures, feeding a pillow which drapes the slope (Figure 21A). The pillow continues growing as long as the lava pressure is great enough to break and stretch its growing and strengthening crust. The crust is simultaneously thickening owing to cooling and thinning owing to stretching as a result of expansion of the pillow. As the pillow grows larger with a constant volume increase, thinning of the crust will decrease more rapidly than thickening because a given increment of volume increase will have less effect on the rate of surface area increase. Eventually the strength of the thickening crust will prevent further growth, and once growth ceases, the contained liquid lava becomes stagnant, thus allowing the crust to thicken so rapidly that it is impossible for the pillow to grow again. The first pillow acts as a barrier to the flow lobe (Figure 21B) which inflates, lifts its roof, and eventually breaks out at a point above the first pillow where resistance is least. Likewise internal pressure within the original flow lobe causes the top crust to fracture and feeds equidimensional pillows which mantle the roof. Pillows continue to grow at the flow front (Figure 21C), effectively damming the main lava flow and causing it to inflate and break out above the pillows and override them with new pillows. Each new pillow is capable of growing only about to the base of the scarp because growth and expansion require in part stretching due to gravity on the steep flow front. These pillows will, of course, branch laterally also, and some may break into fragments that accumulate at the base of the scarp; both these processes will also effectively dam the flow front. In this way the flow front is maintained at a steep angle of equilibrium, and a channel is propagated near the top of the growing pillow pile (Figure 21C). The feeding channels do not solidify rapidly like the elongate pillows because they are generally larger and, more important, they contain flowing lava that continually supplies enough heat to retard growth of the crust. Continued delivery of lava causes the steep pillowed flow front to advance progressively downslope.

Distinctive pillow types occur in special parts of the pillowed flows (Figures 20–22). The top gentle part of the flow is covered with subspherical large pillows and the steep flow front is draped with smaller-diameter elongate pillows. Commonly, aligned pillows atop the flow

27 Bulbous pillows which have grown atop master flow units.

28 Tops of bulbous pillows largely buried by sediment.

29 Large bulbous pillow showing well-developed spreading crack.

30 Bulbous pillows transitional into elongate pillows on flow margin.

have partly collapsed, revealing hollow interiors. Lava evidently drained from these pillows back into the master flow lobe, and groups of aligned hollow pillows mark its path.

In cross section (Figure 21) a pillowed flow apparently consists of three important elements: (1) a master feeding tube sloping gently downward from the vent and occurring near the top of the flow, (2) a series of smaller foreset-bedded elongate pillows draping the flow front and maintaining the same angle of plunge (30–45°) through the whole flow beneath the master tube, and (3) a thin top layer one or two pillows thick of larger subspherical pillows which have grown upward from the master tube. In addition slumping of the flow front produces a pile of rubble at its base (Figure 20), producing a discontinuous layer of rubble beneath the foreset-bedded pillow layer.

In plan the structure of the pillowed flow is complex with the master feeding tube branching toward the distal end of the flow into many smaller feeding tubes. These override previously fed elongate pillows draping the flow front, and the flow advances in a lobate manner.

The master tubes are capable of conducting fluid lava from the vent to the flow front only as long as fresh lava is able to flow through them at a rapid rate. Heat loss is inhibited by the layer of pillow buds which grow upward and mantle the pipes. However, the pipes eventually reach such a length (a few hundred meters) and divide into smaller pipes so that the internal flow is slowed by friction to the point that heat loss to the water exceeds that from the inflowing lava and the pipe freezes shut thus blocking any ingress of fresh lava. When this occurs, lava may break out of the source end of the tube and build a pillowed cone above the vent area. If the supply of lava is sufficient, one or more of the radial elongate lobes of the cone takes over as master feeding tubes of a new flow.

Individual Flow Units

The compound pillowed flows are made up of individual units which flowed and cooled at different times during the development of the pillowed flows. These individual flow units include a great variety of shapes and sizes, some of which are common in particular parts of the master flow units (Figures 20–22). Others provide insight into the processes of pillow formation and the direction of flow. The individual flow units are inter-connected and are generally rounded in form, and range in diameter from a few centimeters to a few meters; they are called *pillows.* The pillow types described here mark a continuum of gradational and combined forms. Commonly several different types are formed in the same place or attached to one another (Figure 22), as is apparent in a study of the photographs. However, an attempt has been made to name and describe the characteristics of each distinctive type so as to better understand their origin.

31 Flattened pillows.

32 Flattened pillows on central part of master pillowed flow, near crest of Mt. Pluto, a volcanic ridge in the center of rift valley.

33 Flattened pillows on top of flow unit.

34 Flattened pillows near vent area on top of volcanic ridge. Note basin-shape of large partly collapsed pillow in center.

Bulbous Pillows

Bulbous pillows are subspherical, high-standing pillows (Figures 23–30) which commonly occur on the top, relatively gentle surfaces of pillowed flows (Figure 20). Such pillows are among the largest individual pillows that were observed in the median valley; they are commonly about 1 meter in diameter and some bulbous pillows 3 or 4 meters in diameter were seen. The bulbous pillows presumably overlie, and were fed from, larger pipe-like tubes. Hence, they grow generally upwards and expand laterally and provide an insulating carapace over the top of the feeding tubes.

Flattened Pillows

Flattened pillows are roughly equidimensional in plan, but present a limp, flattened appearance when compared with the bulbous pillows (Figures 31–35). They are common near the summits of the major volcanic centers such as Mt. Pluto. They are believed to be produced by rapid growth and partial drainage or spreading of bulbous type pillows while still plastic. They appear to mark the sites of eruptive centers where rates of lava delivery are high compared to that of the bulbous pillows. Because of rapid growth, the flattened pillows are confined, by a thinner, weaker solidified crust and tend to flatten and spread under the influence of gravity. Some show distinct central depressions where the partly solidified crust sagged and subsided as the fluid lava interior spread laterally or flowed back into connected distributary lava tubes.

35 Flattened pillows on top of major pillowed flow.

Elongate Pillows

Elongate pillows are sub-cylindrical in shape (Figures 36–46) and where well exposed can be traced through a length 10 to 20 times their diameter. The elongate pillow axes plunge downslope at angles of 10–90° (commonly about 30°) and generally drape the front and sides of a pillowed flow (Figure 20). A sequence of foreset-bedded elongate pillows probably makes up the bulk of most flows, but the top of the master flow may be covered by large equidimensional bulbous pillows. Generally the elongate pillows are difficult to trace for a great distance because they are tangled with, and often covered by, their neighbors.

Elongate pillows are complex in detail. They commonly branch into two or more similar cylinders, they are decorated by smaller pillow buds, and they change direction abruptly depending on the character of the pillowed surface over which they flow while maintaining a general downslope flow direction. They

36 Elongate pillows.

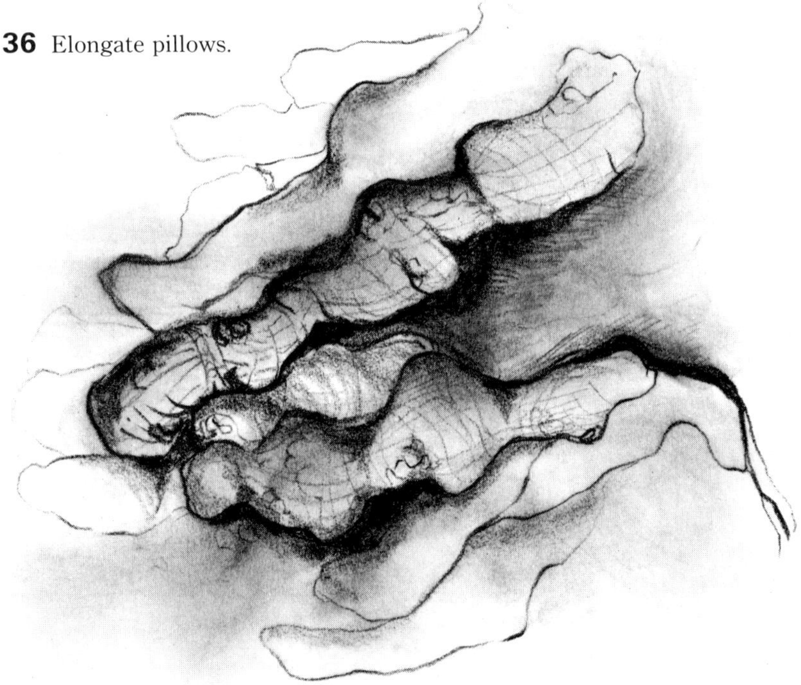

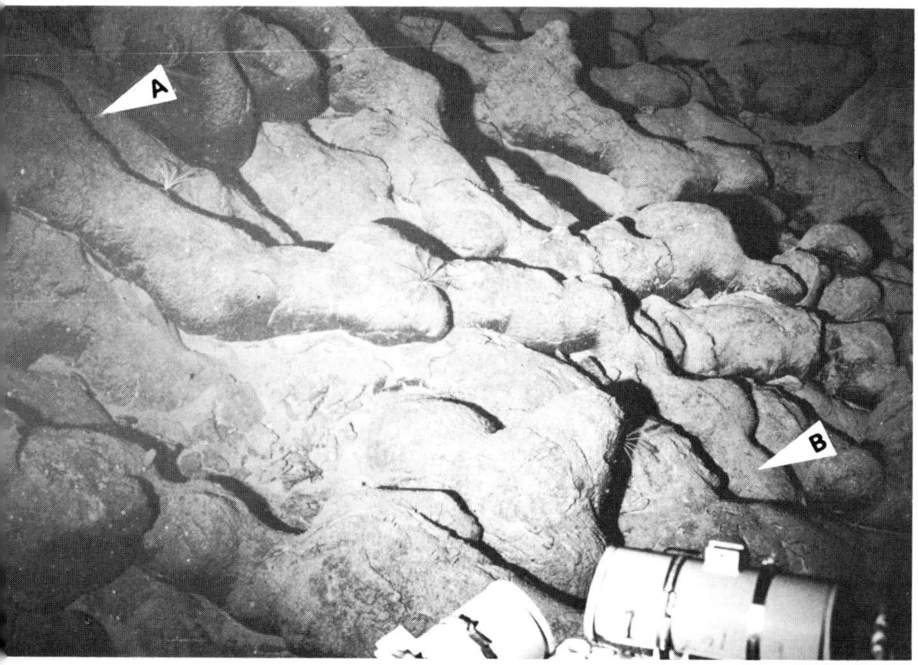

37 Mosaic showing elongate pillows and knobby pillows draping flow front. Note blunt terminations of pillows and overlap of successive younger pillows on upper part of longer, older pillows. Typical spreading crack with matching corrugations shown at A.

38 Elongate pillows draping rather gentle flow front and plunging at 20–30 degrees. Note neck-and-knob shape of tangled pillows and relatively smooth surface texture. One pillow (A–B) can be traced nearly across entire photograph.

commonly develop a neck-and-knob profile (Figure 22) formed by growth from successive downslope-jumping, spreading cracks (Moore, 1975). Elongate pillows that terminate downslope with a distinct hemispherical cap are termed *phallic pillows* (Bellaiche *et al.,* 1974).

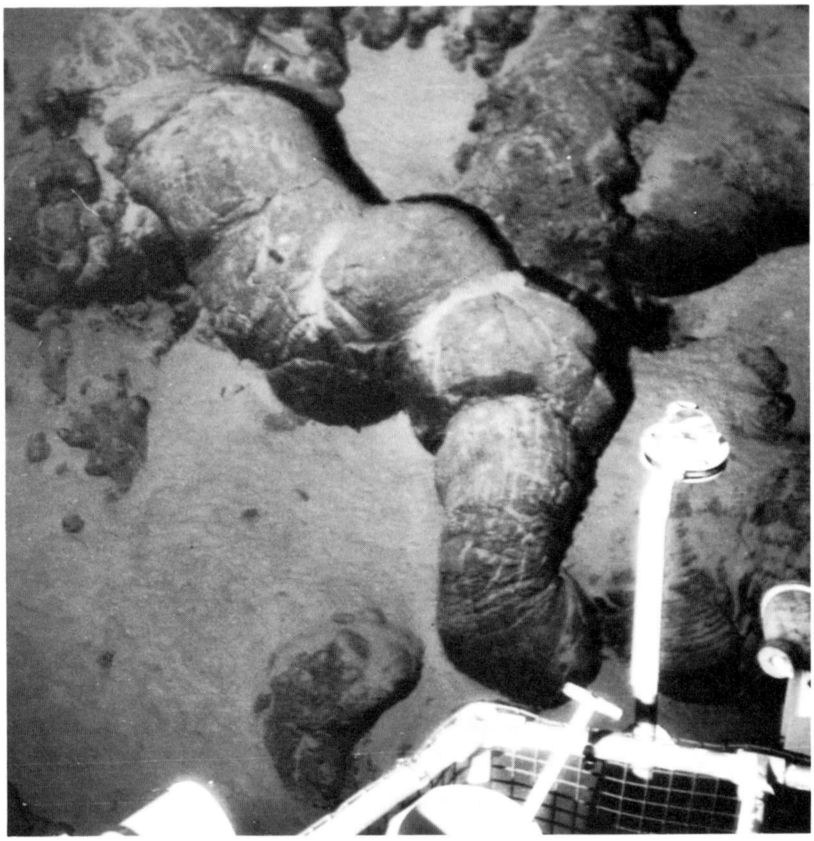

The elongate pillows provide good criteria for determining the direction of flow, and observation of these features can be useful in understanding ancient pillowed flows. In addition to the general plunge in the direction of flow, the elongate pillows branch in the direction of flow. Furthermore, elongate pillows commonly have a blunt or rounded termination in the direction of flow, and some show a distinct decrease in average diameter in the direction of flow.

The flow of lava through elongate pillows appears to be essentially laminar with greatest velocity in the center of the pillow pipe. This differential flow effectively orients elongate phenocrysts parallel with the pillow walls and may cause a concentration of larger phenocrysts in the center of the pillow (Duffield, 1969).

39 Elongate pillow with typical neck-and-knob shape and abrupt changes in direction. Pillow lobe shows general decrease in diameter in downslope flow direction.

40 Elongate pillows on steep flow front. Several pillow lobes have broken and slumped exposing truncated round outlines with radial joints.

41 Photo-mosaic showing elongate pillows on lava flow front on flank of Mt. Jupiter.

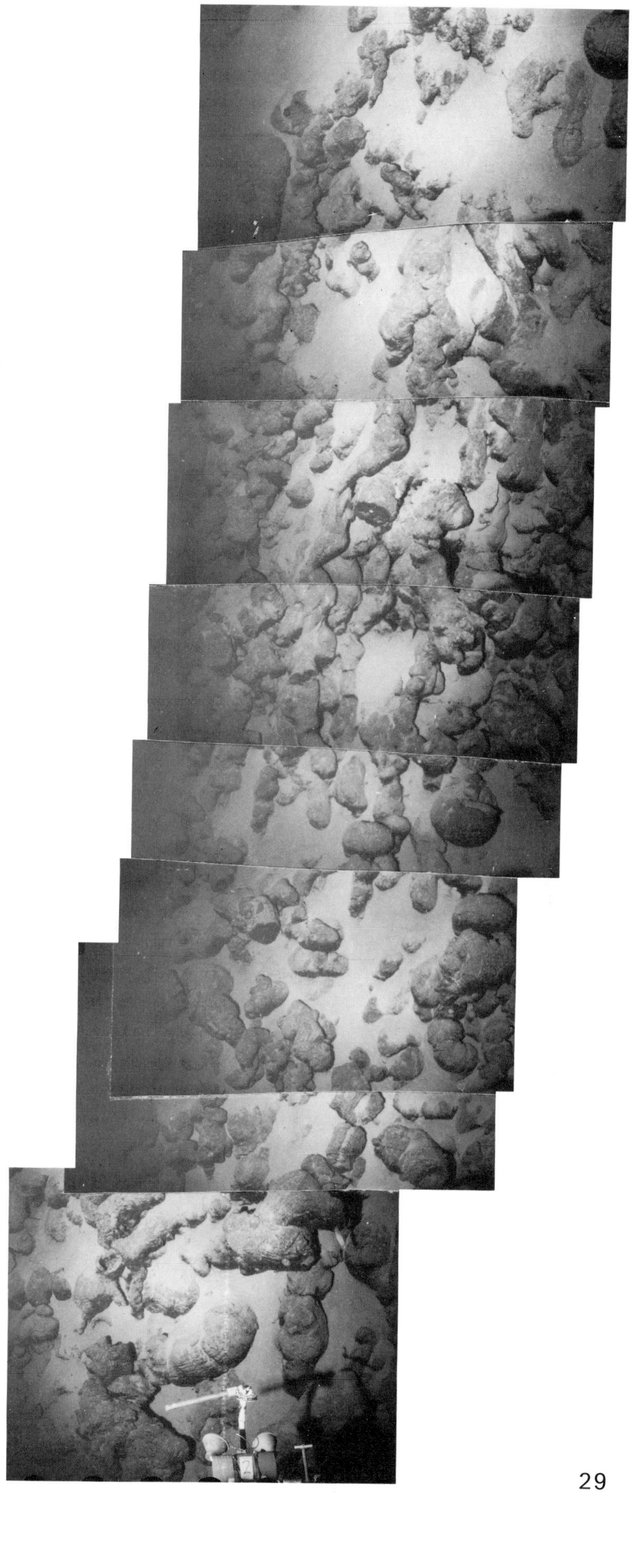

42 Elongate pillows draping flow front.

43 Tangled elongate pillowed flow front. Knobby aspect of elongate pillows makes it difficult to trace the individual flow lobes.

44 Elongate pillow, knobby pillows, and elephant trunks draping flow front.

45 Elongate pillows largely buried in sediment.

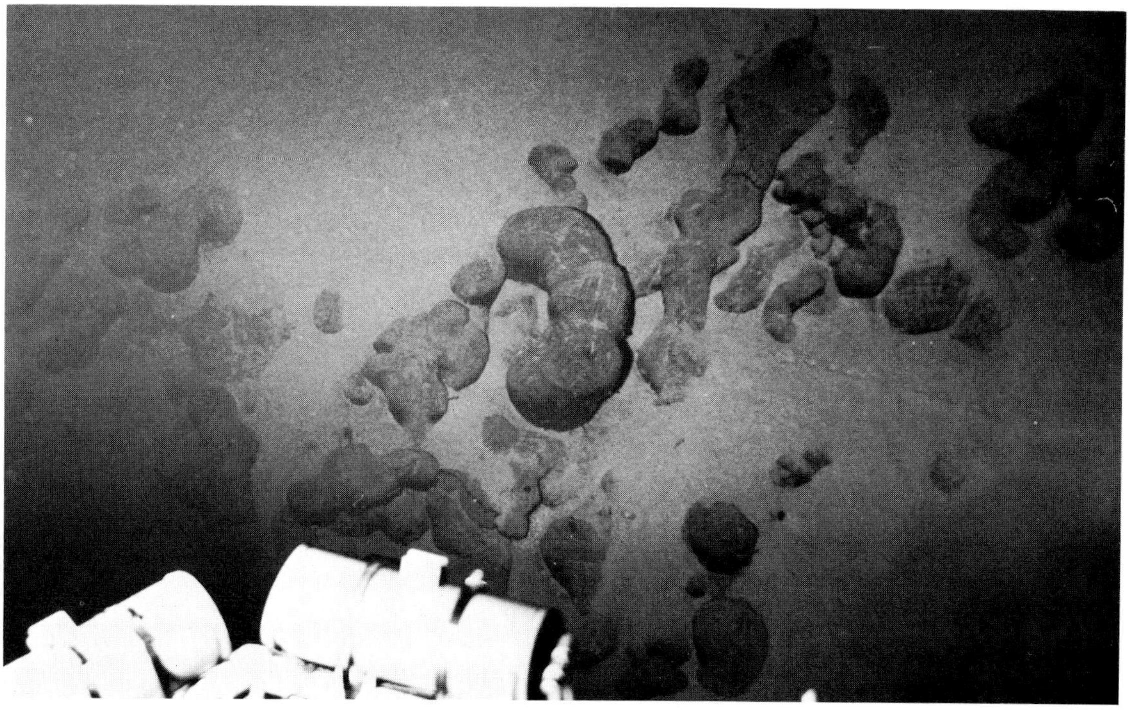

Hollow Pillows

Hollow pillows with an outer shell 1–15 cm thick (averaging 5 cm) are common in a variety of pillow types (Figures 47–64). Hollow pillows are obvious only when broken open; consequently, unbroken ones are unrecognizable. Generally the crust of hollow pillows has broken and collapsed on radial cooling cracks, and the breakage was later than the cooling and may even be unrelated to the volcanic episode that formed the pillow.

The hollow pillows apparently formed when lava feeding a flow lobe, which had chilled and solidified on its outer surface, was suddenly cut off at its source, and the molten lava continued draining downslope through the interior of the tongue (Moore *et al.,* 1973). The near absence of a gas phase at oceanic depths requires that molten lava is replaced by water which is sucked into the hollow pillow pipe through cooling cracks in the outer shell. Commonly hollow pillows occur in a linear group on the upper surface of a pillowed flow (Figure 50). The locus of such hollow pillows defines the course of the feeding underlying tube from which each of the pillows grew and into which lava later drained.

Some hollow pillows, called *hollow layered pillows,* contain subhorizontal lava shelves which mark temporary stillstands of the surface of the molten lava stream flowing through a pillow tube system which is partly full (Figures 48 and 49). As the lava supply successively diminishes, the lava level periodically falls and a new, lower shelf is formed as water enters through cracks in the top crust (as well as through cracks in the upper shelves). Repeated fall of the fluid surface may produce as many as 6 or 8 parallel, subhorizontal shelves separated by flat cavities within the pillow system.

The bottom of the upper crust of the hollow pillow, as well as of each shelf may be decorated by thin, down-hanging septa and stalactites. The curtain-like septa are commonly 1 cm thick and as

46 Elongate pillows on flow front some distance from camera.

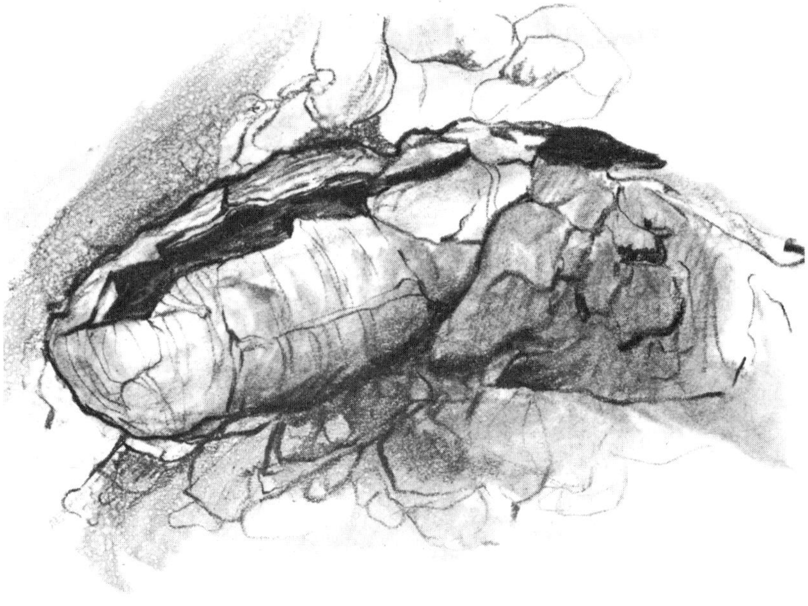

47 Hollow pillow broken open at top showing internal cavity.

48 Broken hollow layered pillow with several internal thin, subhorizontal shelves, each marking the top surface of the fluid lava previously partly filling the pillow.

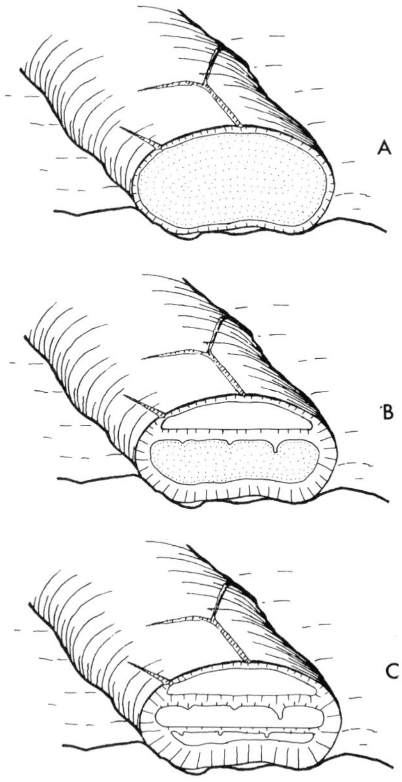

49 Sketch showing development of shelves in a hollow layered pillow. A. Elongate pillow pipe with thin outer crust filled with molten lava (dotted). B. Reduction in volume of flow causes lava level to fall and water to enter upper compartment through cracks in crust where it chills top of fluid lava stream. C. Lava supply is totally cut off after hesitating momentarily and producing a second shelf; fluid lava drains downslope emptying tube system at this level. Note septa hanging down from shelves and progressively greater wall thickness downward. In addition, glass thickness is generally greater on top than on bottom of shelves.

50 Vertical LIBEC photograph showing three aligned collapsed hollow pillows 1.5 to 3 m in diameter. Narrow dimension of photograph about 17 m.

much as 10 cm wide and the stalactites may be about 1 cm in diameter at their point of attachment. Both commonly hang downwards several centimeters.

These features all aid in determining the original orientation of an unattached fragment of a hollow pillow. The lava shelves are almost horizontal; their slight downstream slope is generally less than that of the pillow lobe in which they occur (Waters, 1960). The side walls of a hollow layered pillow are progressively thicker downward where they have had a longer history of solidification (Figure 49). The top of the shelves is flatter, with a thicker glass rind because it has been more drastically quenched, and the bottom may be decorated by septa and stalactites.

Hollow pillows that break open when the crust is still very thin and plastic must form where lava delivery rates are especially high and erratic. Such *plastic hollow pillows* (Figures 55, 57, 59, 64) commonly tear open on one side and rapidly dump their content of molten lava producing a sheet flow (*broken egg structure* of Bellaiche *et al.,* 1974). Other plastic hollow pillows appear to drain more slowly and feed typical elongate or knobby pillows.

Hollow pillows and their internal shelves are fragile and easily broken. Consequently, they readily provide bite-sized fragments to a dredge dragged over pillowed terrain, and commonly comprise a disproportionately high fraction of dredged samples.

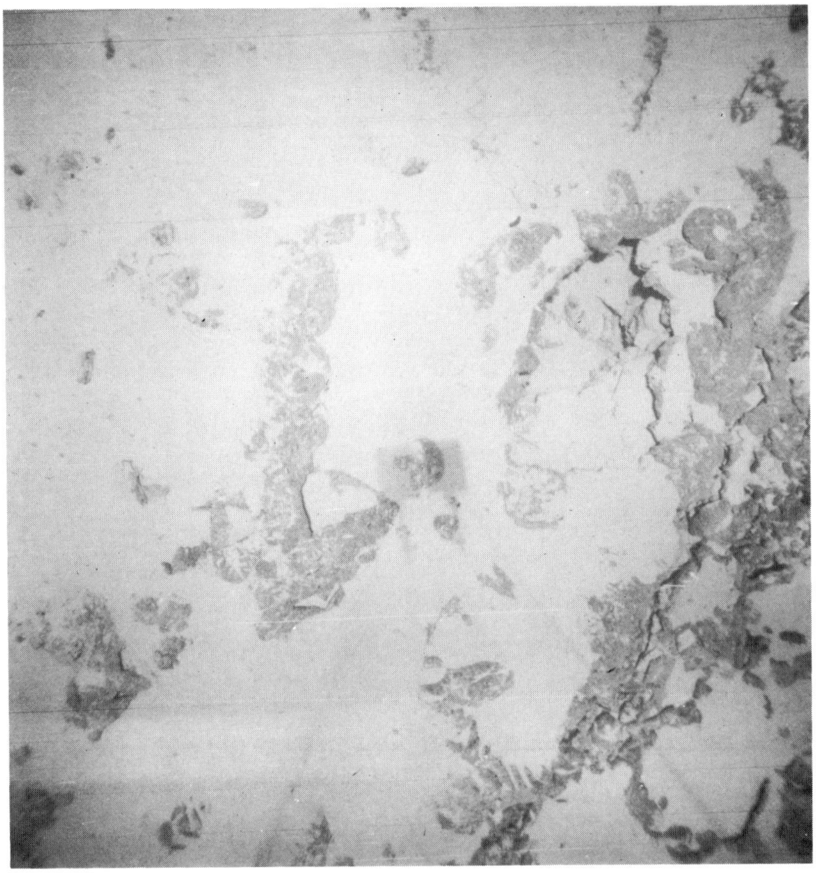

51 Vertical LIBEC photograph showing large collapsed hollow pillows (right center) about 3 m long. Narrow dimension of photograph about 14.5 m.

52 Hollow pillow broken open and partly filled with sediment.

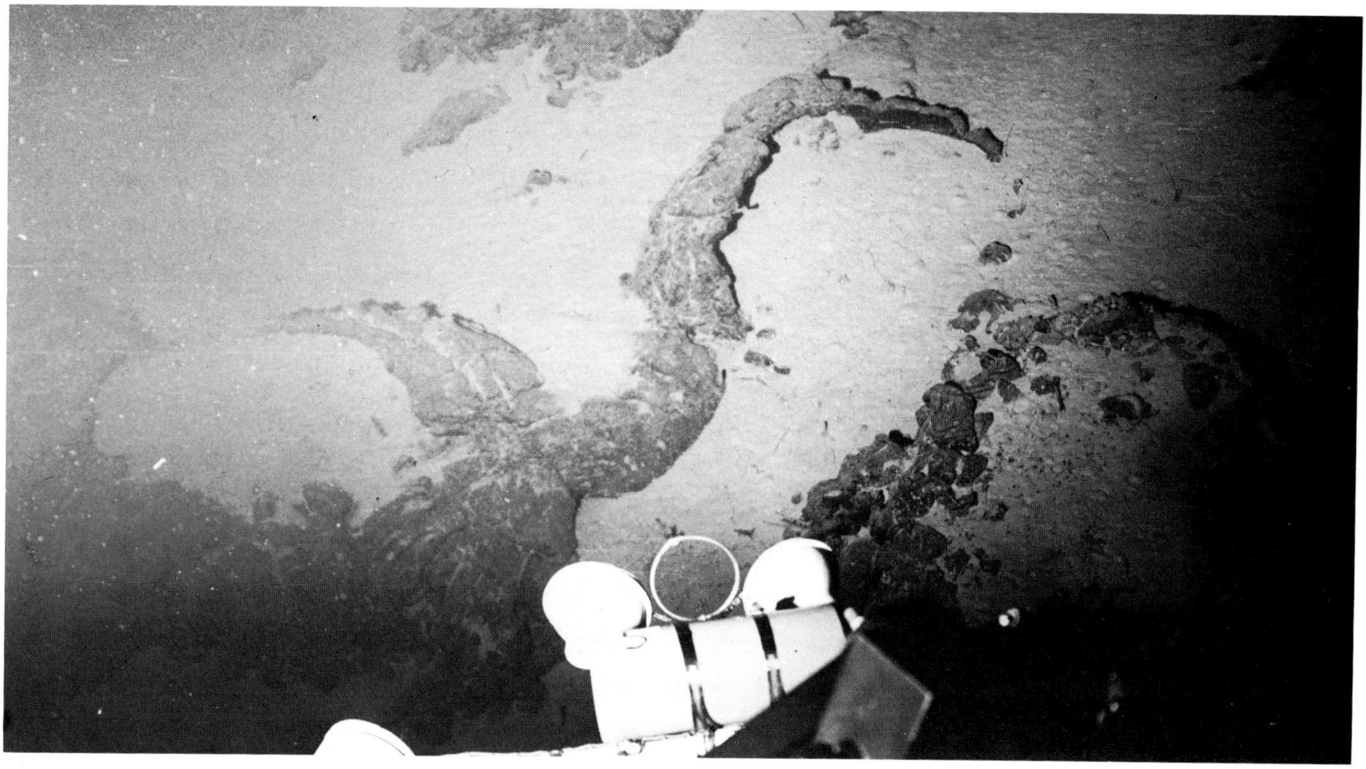

53 Group of three sediment-filled, hollow, bulbous pillows on flow top. Pillows are probably connected to same feeding tube system.

54 Broken hollow pillow containing curved internal shelves.

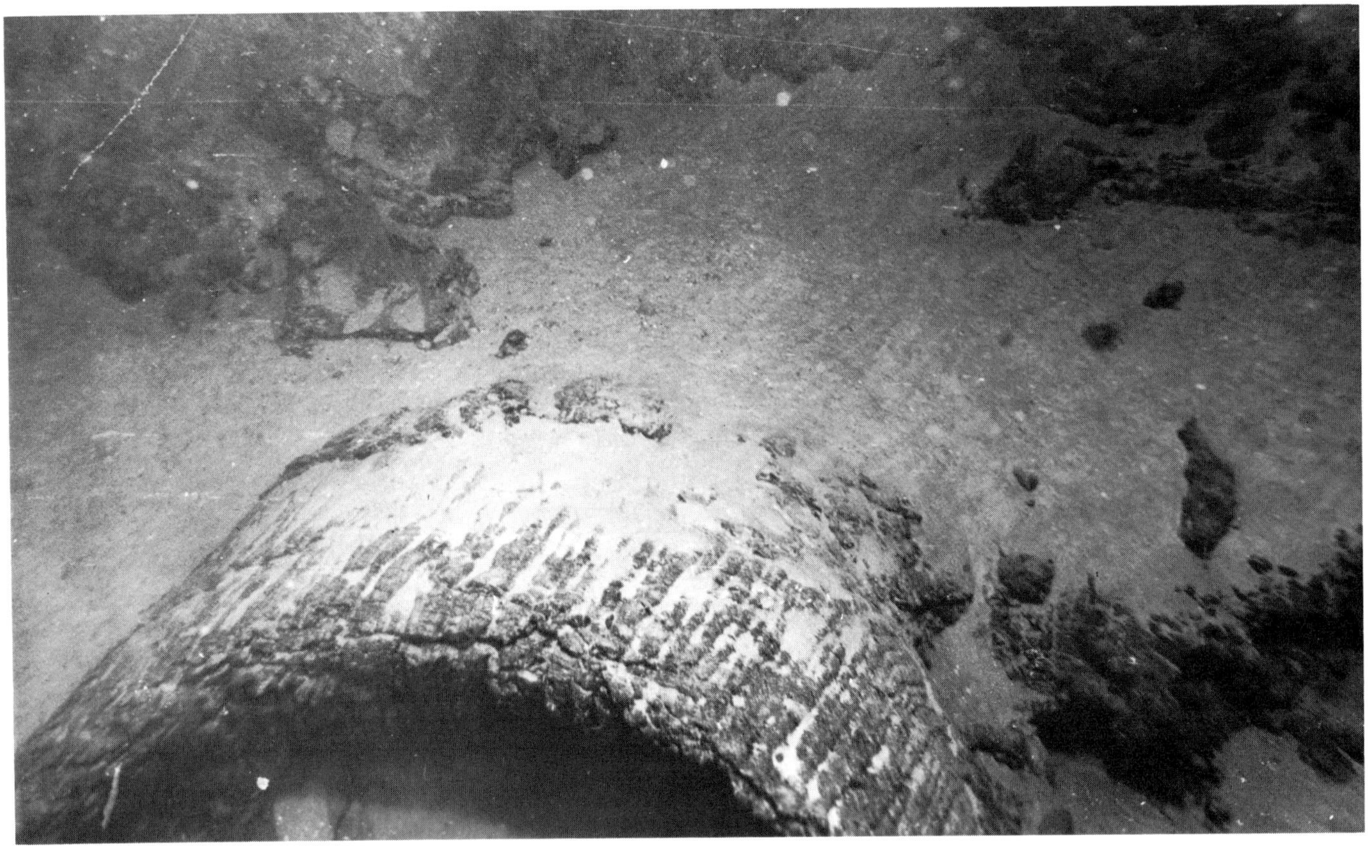

55 Plastic hollow pillow which apparently broke open and drained while outer crust was still partly plastic.

56 Large, sediment-covered bulbous pillow on flow top. Small hole indicates that pillow is hollow.

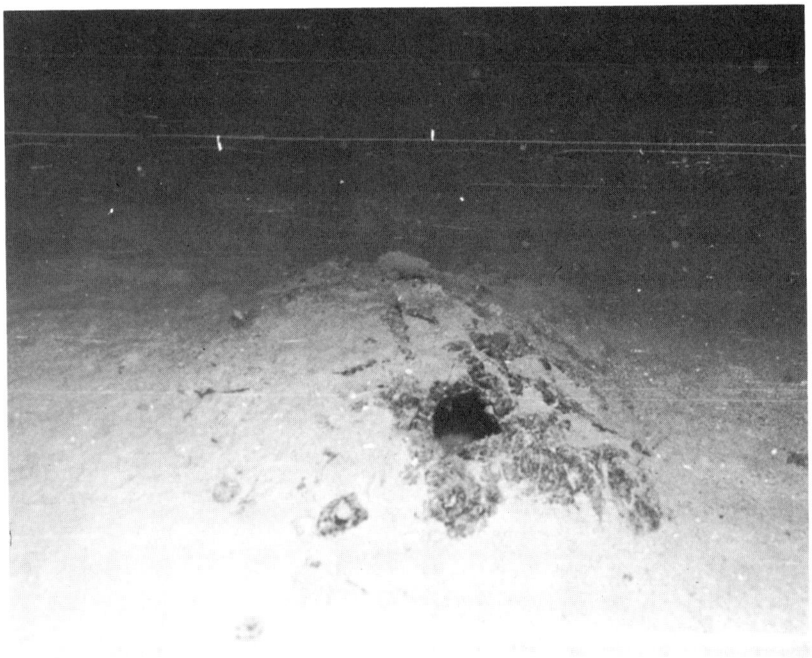

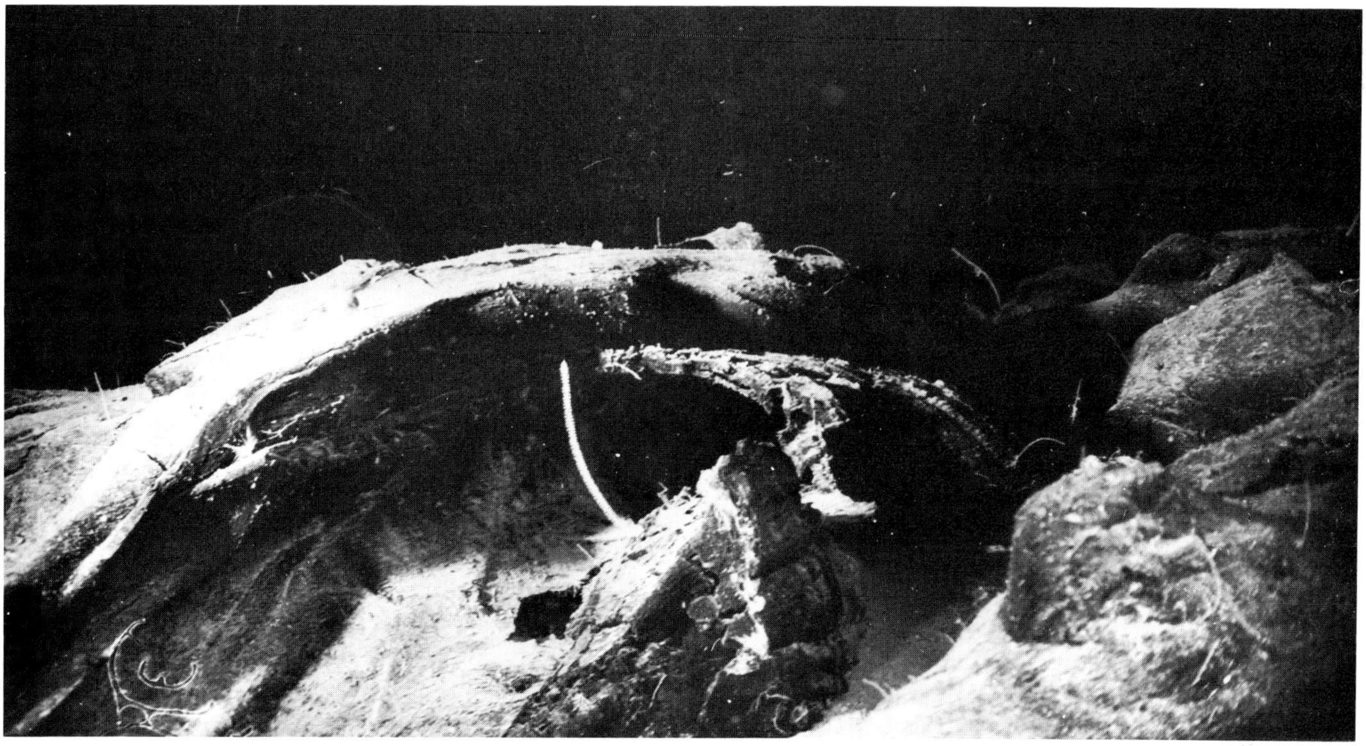

57 Plastic hollow pillow with very thin crust which drained and collapsed shortly after forming. Photograph taken from French submersible CYANA.

58 Small broken hollow pillow exposed in fault scarp (center right) showing three temporary still-stands of molten lava within the pillow system.

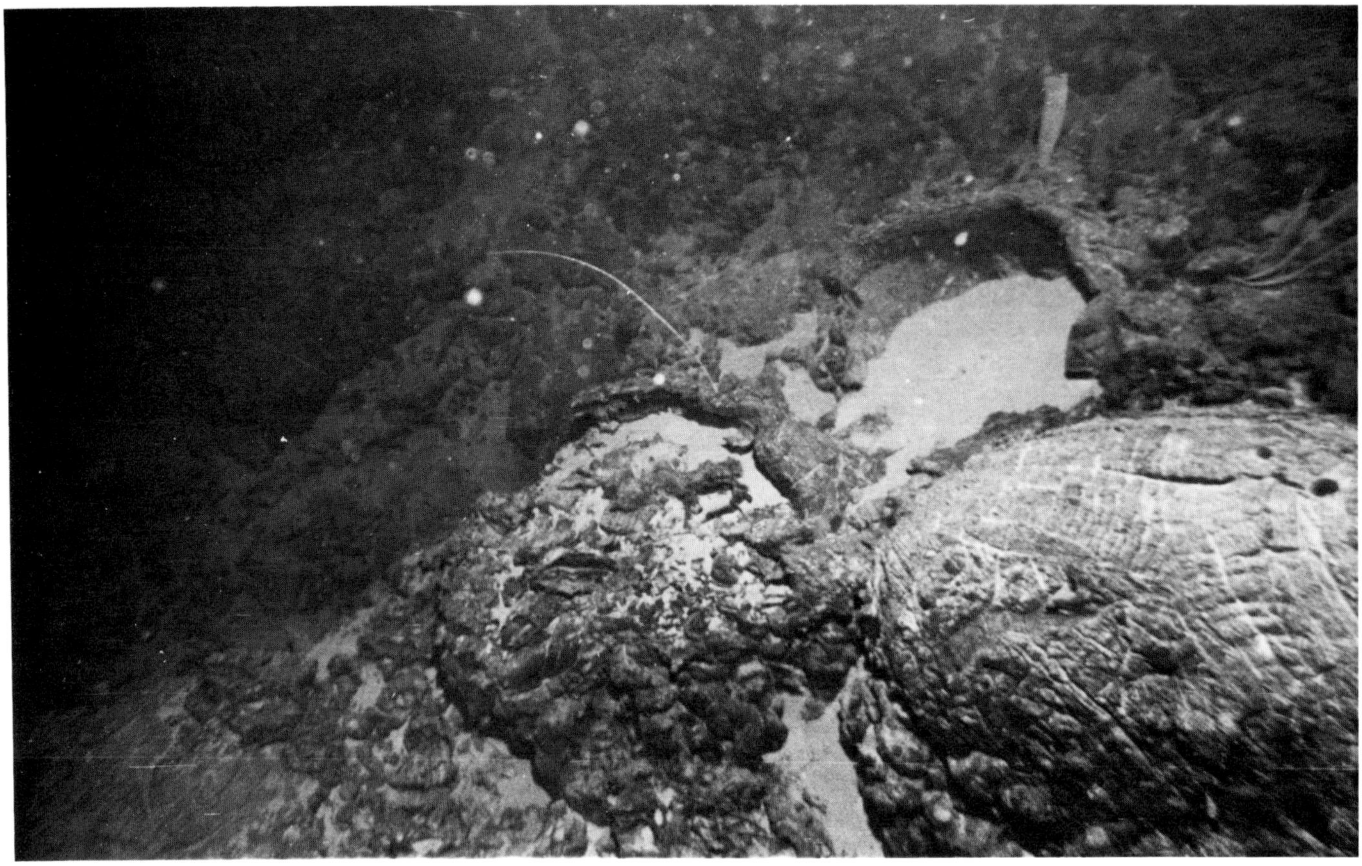

59 Two interconnected plastic hollow pillows on flow front. Irregular knobby pillow is fed from contents of lower partly hollow pillow.

60 Broken, hollow pillow largely drowned in sediment.

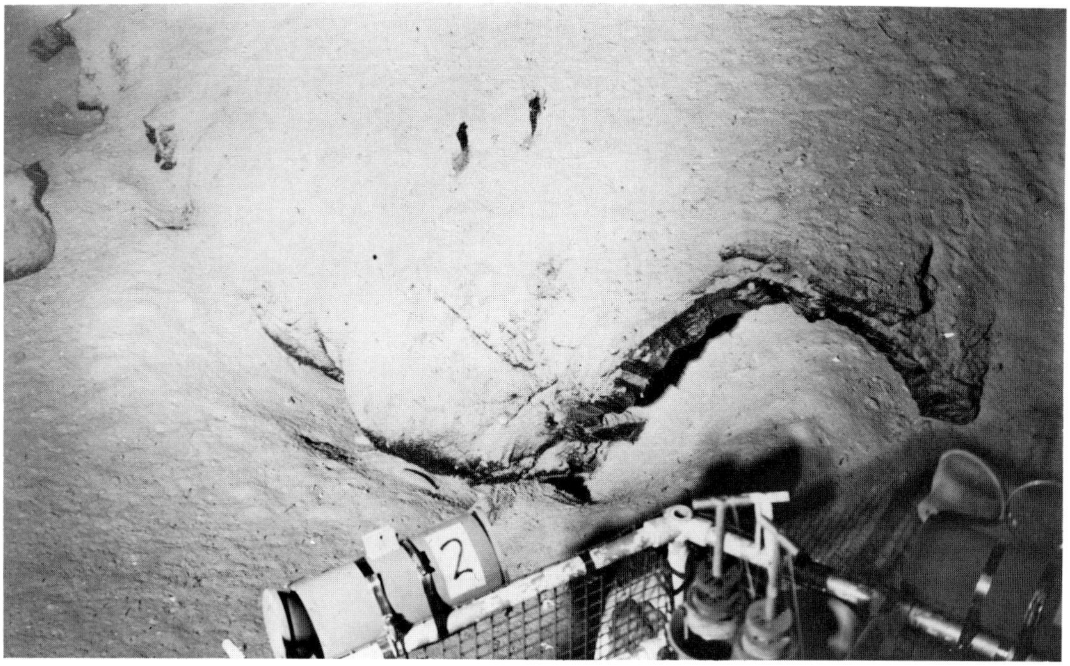

61 Sediment covered hollow pillow. Sediment has sifted down into two small cracks on top surface.

62 Hollow, layered pillow partly filled by sediment. One partly collapsed shelf is visible.

63 Hollow, layered pillow containing a large number of thin subhorizontal shelves, each representing a temporary still-stand of liquid lava filling the pillow system.

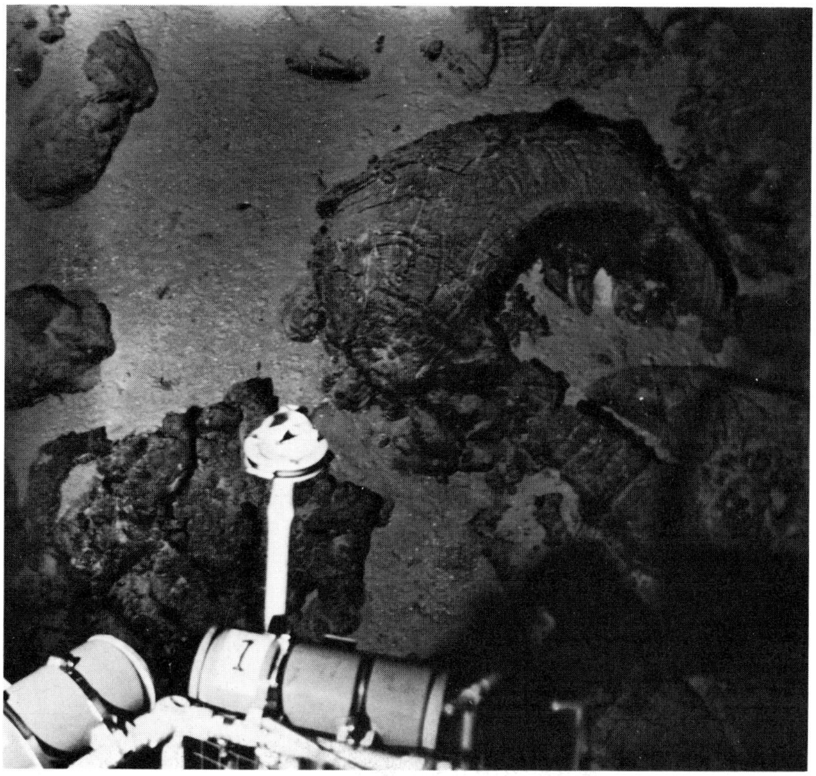

64 Partly collapsed, hollow plastic pillow.

Knobby Pillows

Knobby pillows are either bulbous or elongate pillows which are decorated by many small pillow buds (Figures 65–75). The individual buds are 2–10 cm in diameter, have a rounded end, and are commonly 2–4 times longer than their diameter. They are generally most abundant on the lower half of the parent pillow from which they grow, but such buds are not uncommon growing from the top of bulbous pillows. The pillow buds have variously been called *pillow fingers, elephant trunks,* and *tripe* (Belaiche *et al.,* 1974). The mini-pillows described in ancient pillowed flows (Dietrich, 1967) may be pillow buds of this type.

Because of their small size, the buds of knobby pillows undergo particularly rapid cooling and consequently develop glassy crusts thicker than that which forms on the larger pillow forms. When the pillow fingers are less than 4 or 5 cm in diameter, they are composed almost entirely of glass with no microcrystalline core. This form of cooling, therefore, produces the largest masses of nearly pure basalt glass, particularly in basalt types free of phenocrysts,

65 Knobby pillow.

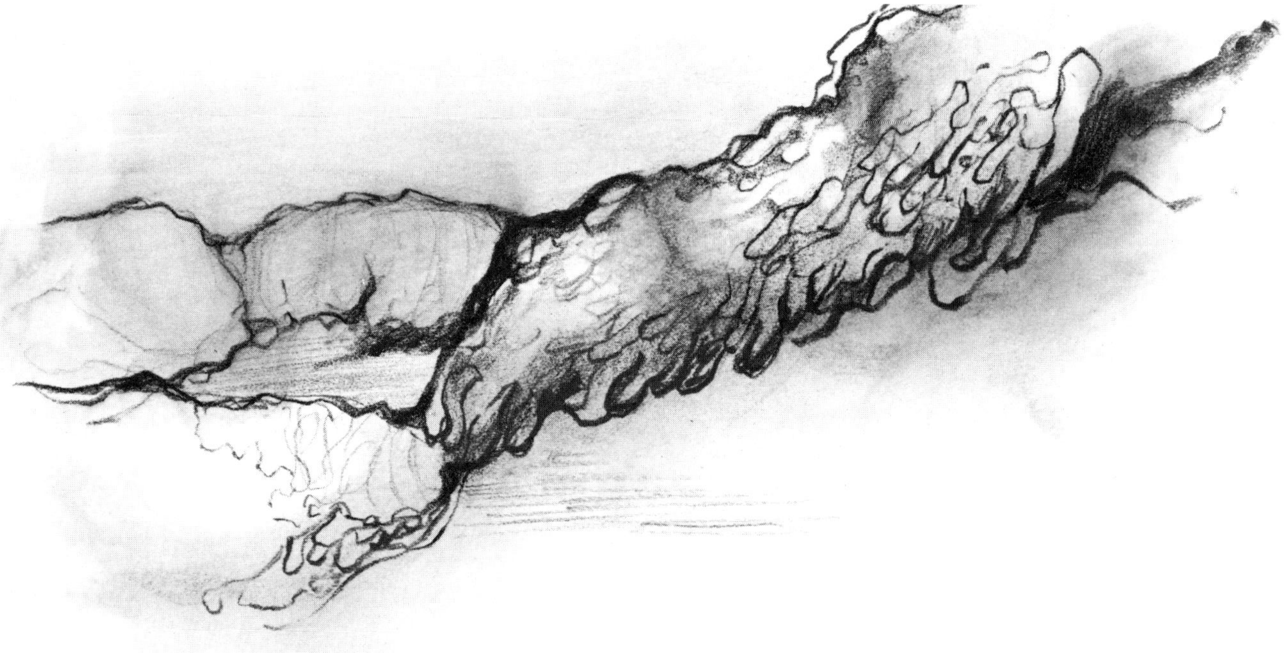

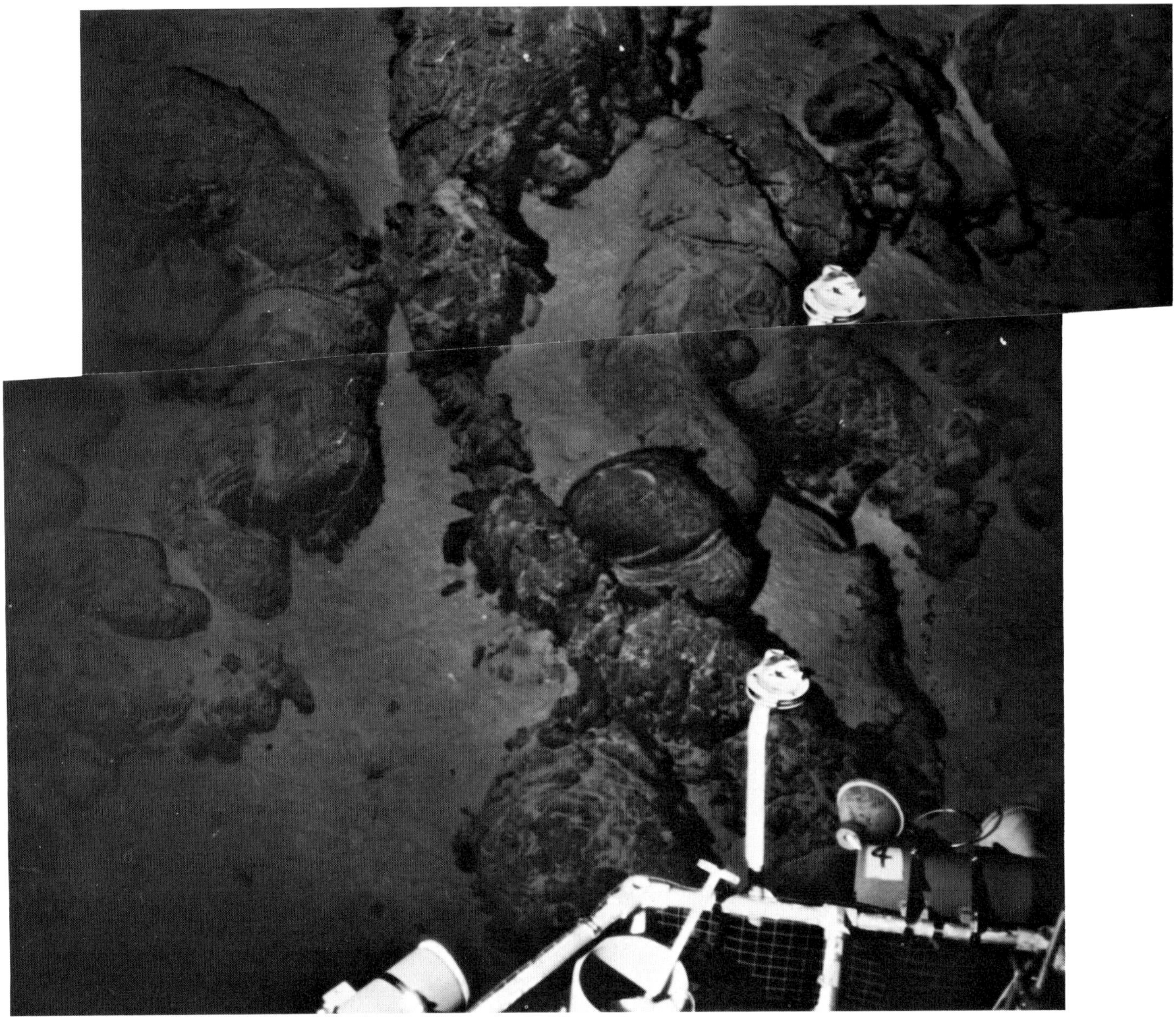

66 Mosaic of two photos showing elongate knobby pillow draping sedimented flow front. Flow lobe clearly branches into two knobby pillows at lower part of photograph.

which is strongly stressed due to its rapid chilling and contraction. Raising of such material to the sea surface, where the temperature is higher, and the pressure lower, causes some specimens to spontaneously spall and explode on the deck of the ship as the so-called *popcorn rocks* (Hekinian *et al.,* 1973). Commonly such pillow fingers will continue to crack and spall for months and years after collection, particularly if roughly handled.

Knobby pillows appear to have grown rapidly where flow rates are high, particularly on steep flow fronts. They mark

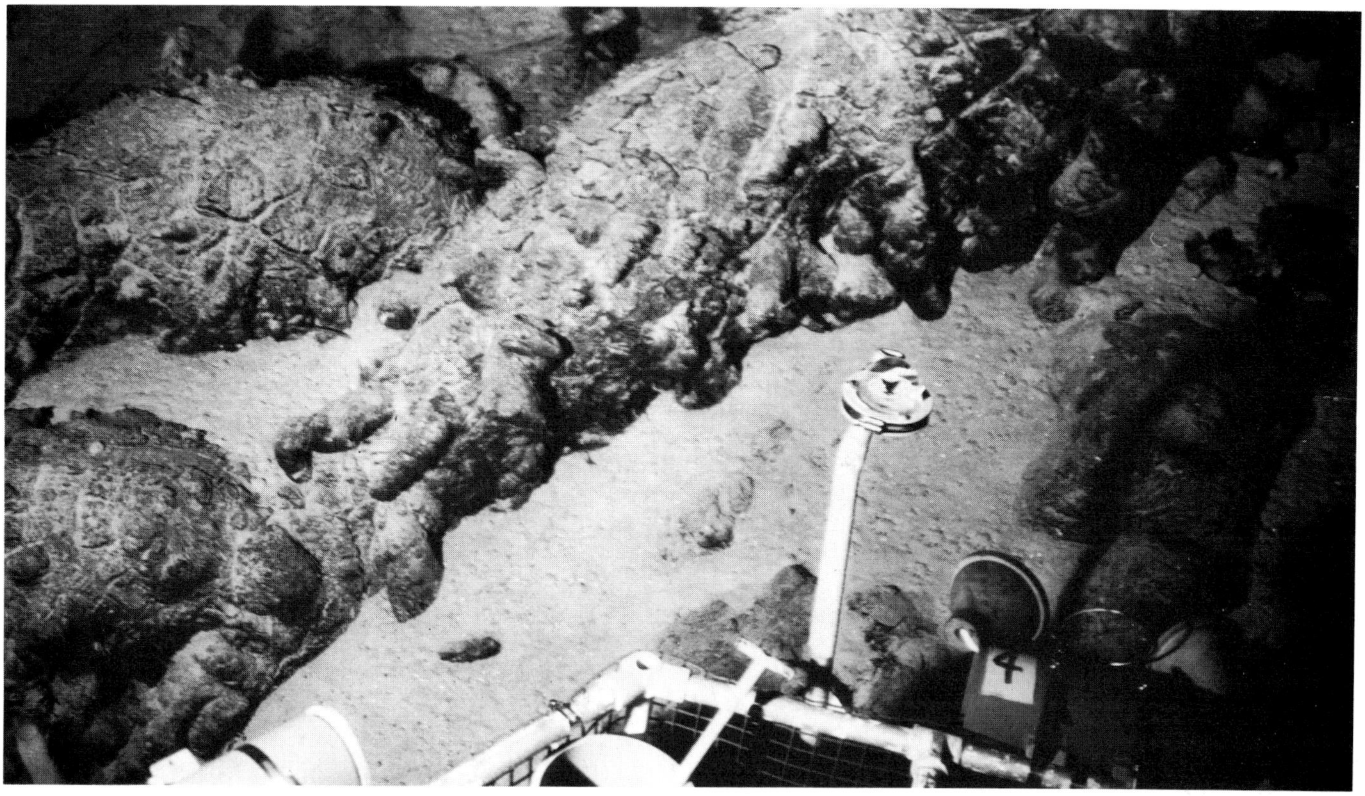

67 Knobby pillow on slope. Note how pillow fingers generally occur on lower half of larger pillows and terminate in downflow direction toward lower left. Note breadcrust cracks in pillow on left.

68 Mosaic of knobby pillow with many small pillow buds, some growing straight up. Small bud at base of pillow on left side has clearly grown from vertical crack. Note that some small pillow buds have broken off and lie on top of sediment.

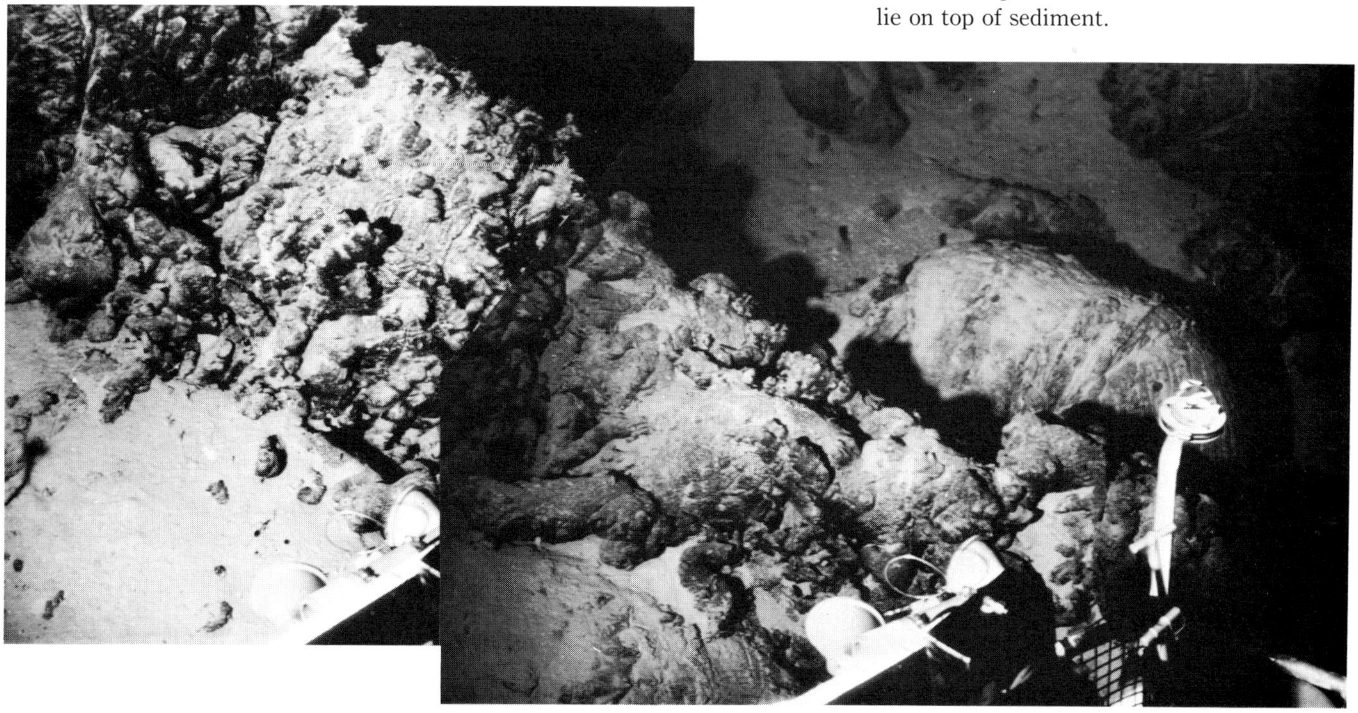

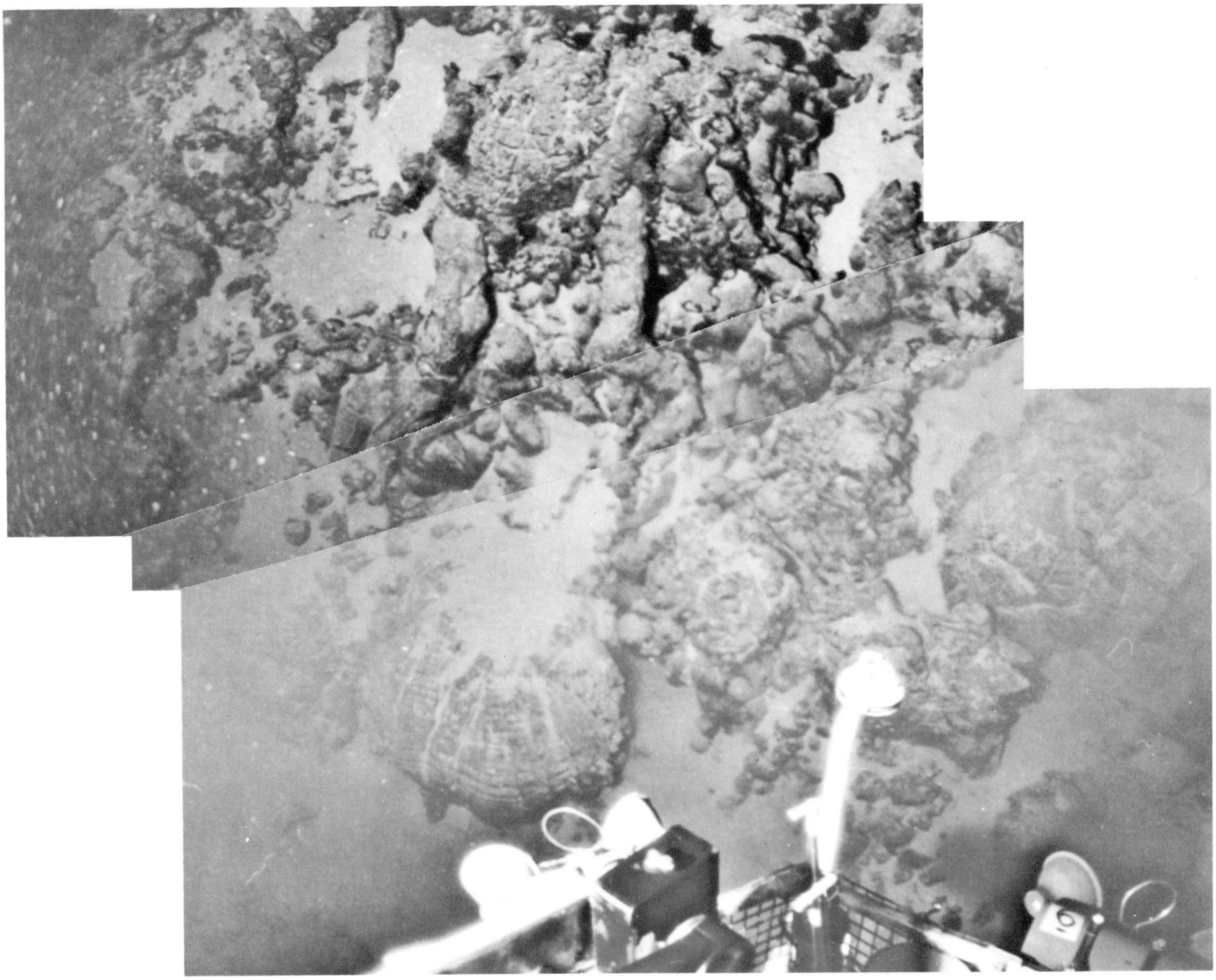

69 Mosaic of three photos of knobby pillows draping flow front. Knobby mass on left terminates on bottom in typical bulbous pillow with several radial spreading cracks.

an intermediate type of cooling unit between normal ridged pillows (slow growth) and sheet flows (very rapid delivery of lava). The knobby pillow has grown slowly enough so that a large pillow form can develop because of containment within its outer crust, but rapidly enough so that this crust can rupture in many places and feed a large number of small pillow buds and fingers. Although these buds begin growth in any direction, they hang down if they grow very long, and those that grow from near the top of a bulbous pillow may be almost vertical. This fact and the general downslope direction of shorter buds can provide evidence on the general slope, and direction of flow of a sequence containing knobby pillows.

70 Knobby pillows including elongate fingers hanging down from their point of attachment. Upward-growing siliceous sponges in background indicate vertical direction.

71 Knobby pillows draping flow front. Note how largest pillow fingers break out of lower part of parent pillows. Continued expansion of one of these fingers could lengthen and expand entire pillow system.

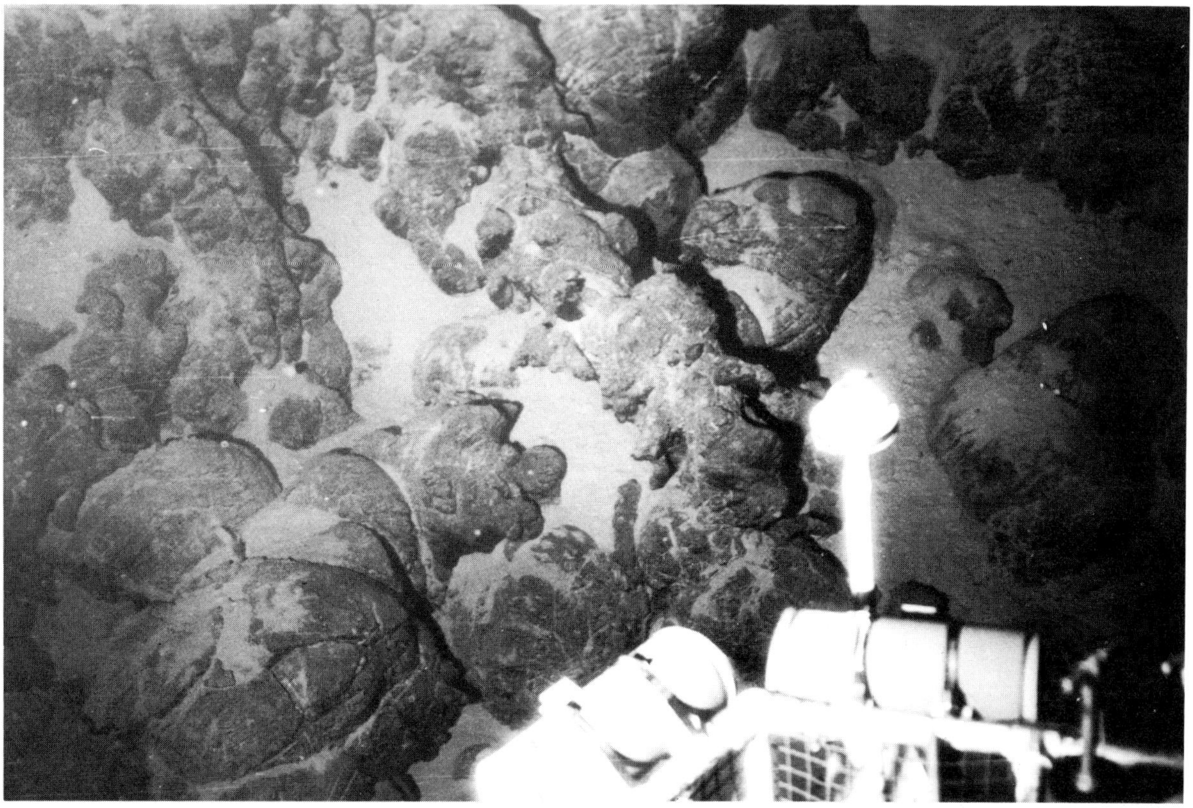

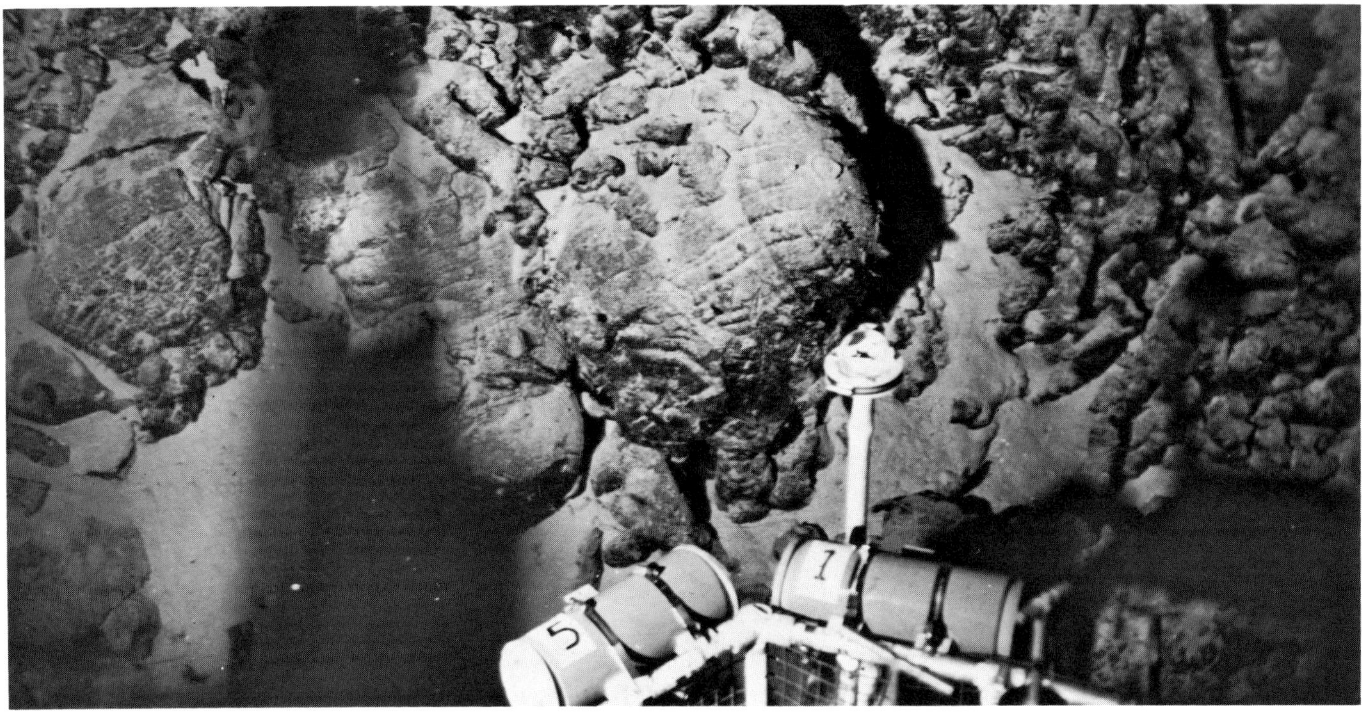

72 Partly sediment-covered flow front draped with knobby pillows.

73 Pillow buds and elephant trunks growing from bulbous pillows on flow front.

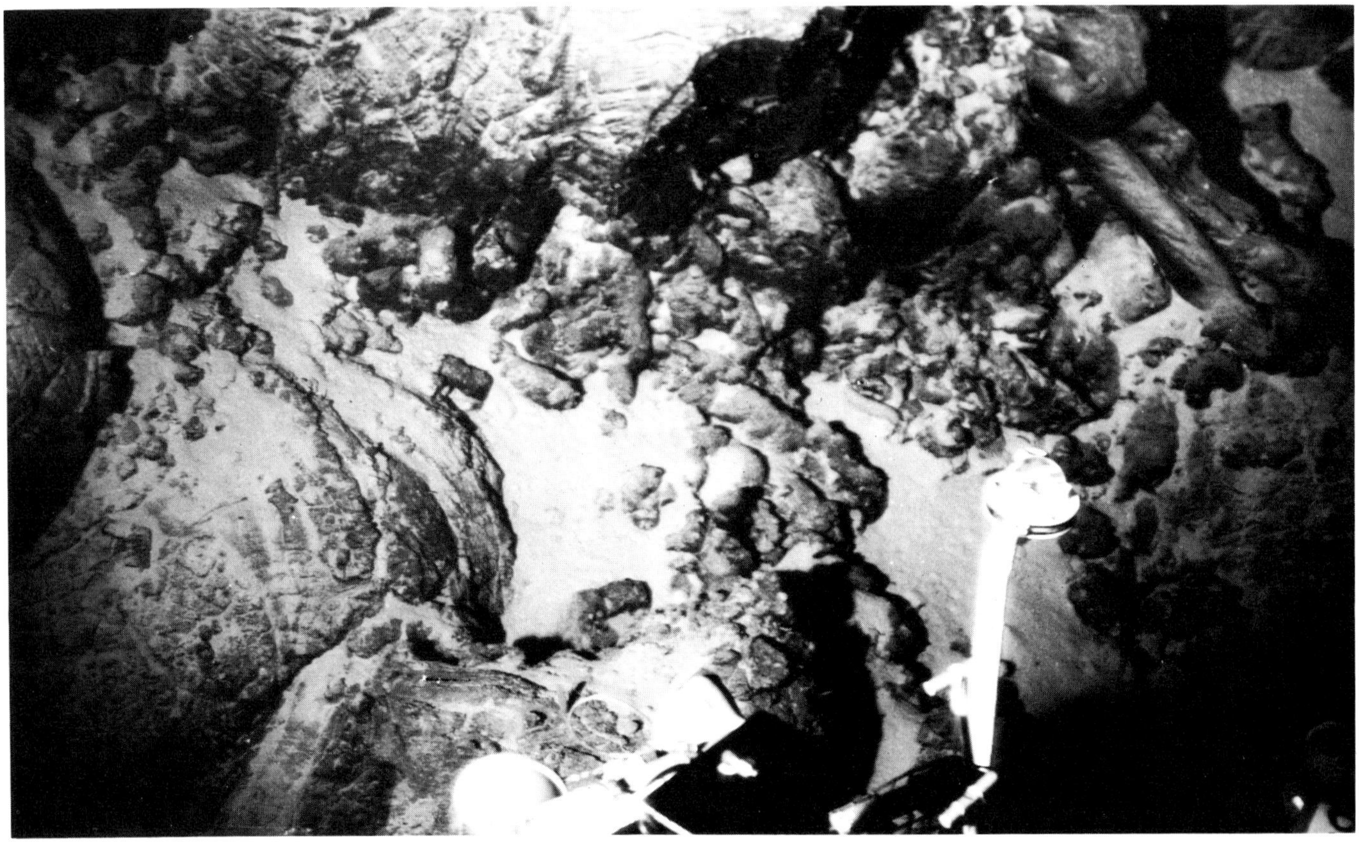

74 Knobby pillows decorated with a variety of pillow buds partly buried in sediment.

75 Knobby pillows with well-developed elephant trunk pillow buds.

Trapdoor Pillows

Trapdoor pillows are a specialized variety of pillow bud which grow from a circular crack cutting the surface of a larger pillow (Figures 76–84). They are short and stubby, having grown only for a short time, but are of interest because they display well a mechanism by which one pillow lobe splits off from another (Figure 22). The crust of the parent pillow caps the end of the trapdoor pillow, and the sides are marked by striations or corrugations parallel to the direction of movement of the uplifted cap. The crust of the side of the trapdoor pillow is formed at the circular crack in the original parent pillow (Moore, 1975).

76 Trapdoor pillow.

77 Toothpaste pillow.

78 Two trapdoor pillows (left) on top edge of pillowed flow.

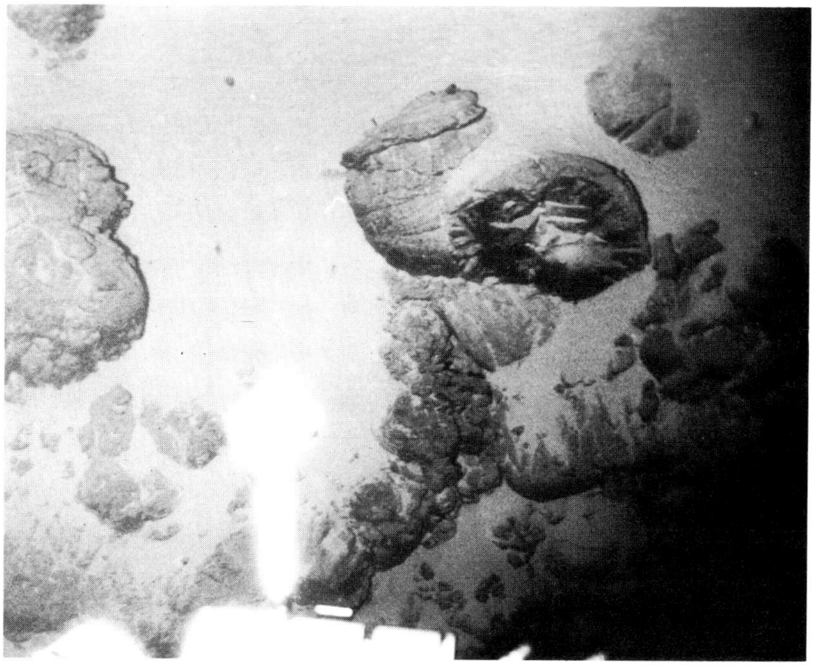

Spreading on this crack may be asymmetrical in which case the crack remains at its original site on the parent pillow, or symmetrical in which case the spreading crack migrates outward and encircles the trapdoor pillow half way from the parent pillow to the end of the trapdoor pillow.

Pillows of this general type which are more elongated and commonly bent are termed *toothpaste pillows* (Figure 77). They form when the hydraulic pressure of molten lava within a confined pillow tube breaks out on a circular crack in the same fashion as in a trapdoor pillow. The pillow bud is pushed upward or sideways, but eventually bends over and hangs downward as its largely plastic core cannot withstand the effect of gravity.

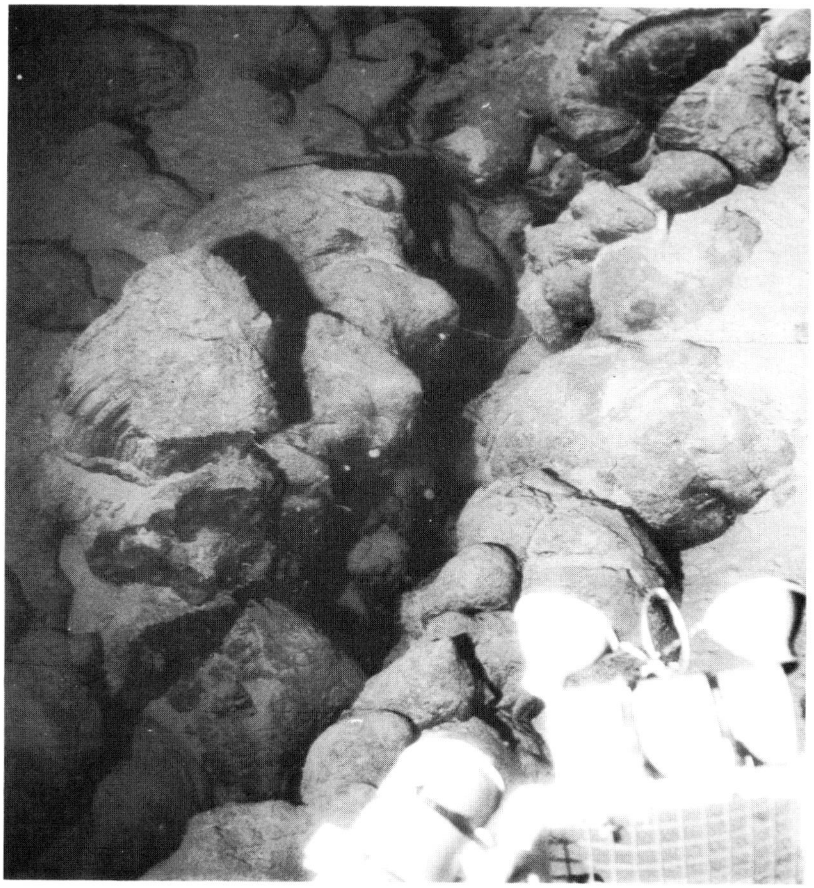

79 Trapdoor pillow (upper right) on sedimented flow front. To the right of trapdoor pillow is broken hollow layered pillow with two subhorizontal shelves.

80 Trapdoor pillow (center left) that grew from pillowed flow surface.

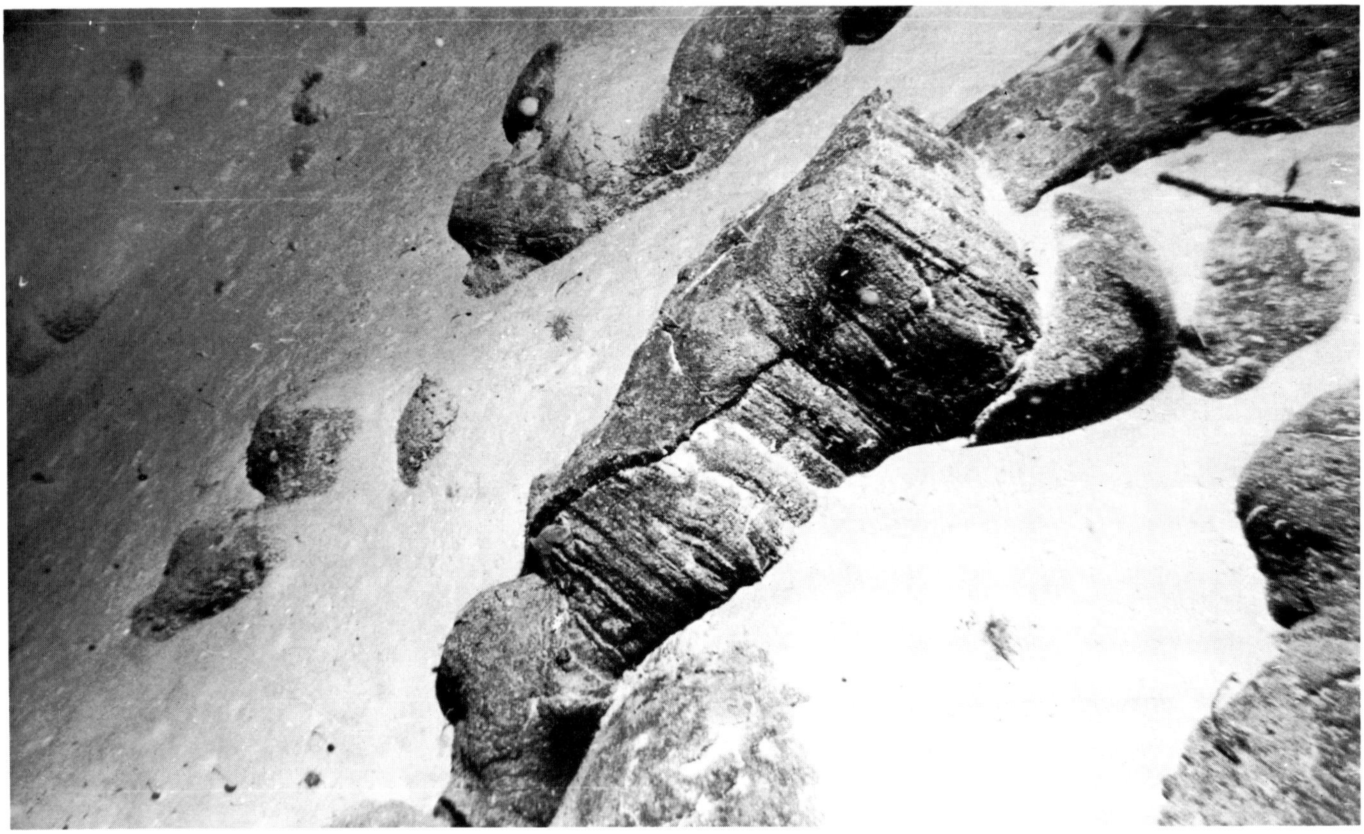

81 Trapdoor pillow on sedimented flow front. Original crust of parent pillow is clearly visible on top.

82 Small trapdoor pillow atop bulbous pillow buried in sediment.

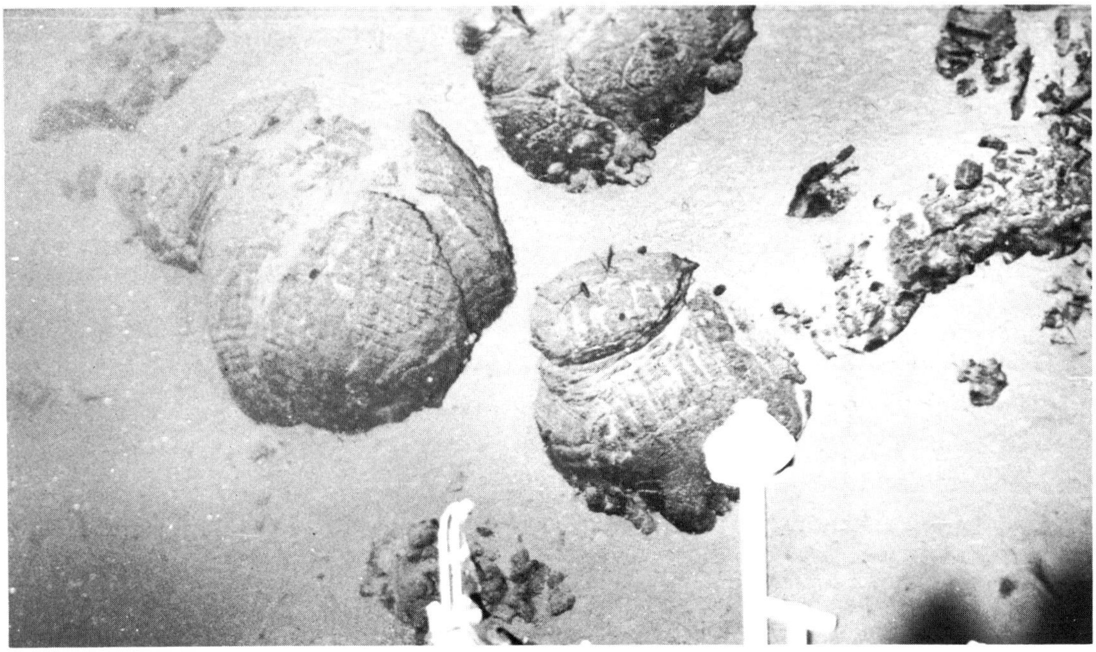

83 Trapdoor pillow hanging down to left. Original pillow crust visible on cap of pillow.

84 Trapdoor pillow which has grown from top of large bulbous pillow.

Sheet Flows

Sheet flows are nonpillowed flows which form where extremely rapid delivery of fluid lava, generally on a steep slope, permits the lava to flow and spread before it can be contained within an outer solidifying skin (Figures 85–89). Sheet flows are generally thin (approximately 30 cm) and commonly display a ropy or wrinkled top surface. They can continue downslope and feed an elongate pillow, commonly a knobby pillow.

A common local source for a sheet flow is the rapid emptying of the molten interior of a hollow pillow by rupturing or plastic failure of its crust.

Sheet flows may also be fed by extremely high eruption rates from primary vents. Dredge hauls (Larson, 1971) and photographs (Normark, 1976) suggest that sheet flows may be common on the East Pacific Rise near the mouth of the Gulf of California.

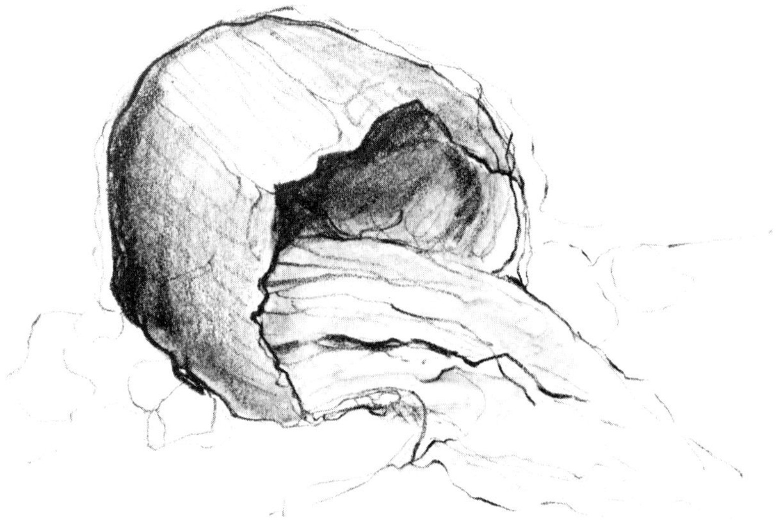

85 Sheet flow fed by draining of hollow plastic pillow.

86 Sheet flow which has poured from a collapsed partly plastic hollow pillow. A second drained hollow plastic pillow is visible on left. Photograph taken from French submersible ARCHIMEDE on dive no. 73-10.

87 Mosaic of 2 photos showing small sheet flow draping flow front and apparently feeding collapsed, hollow, plastic pillow.

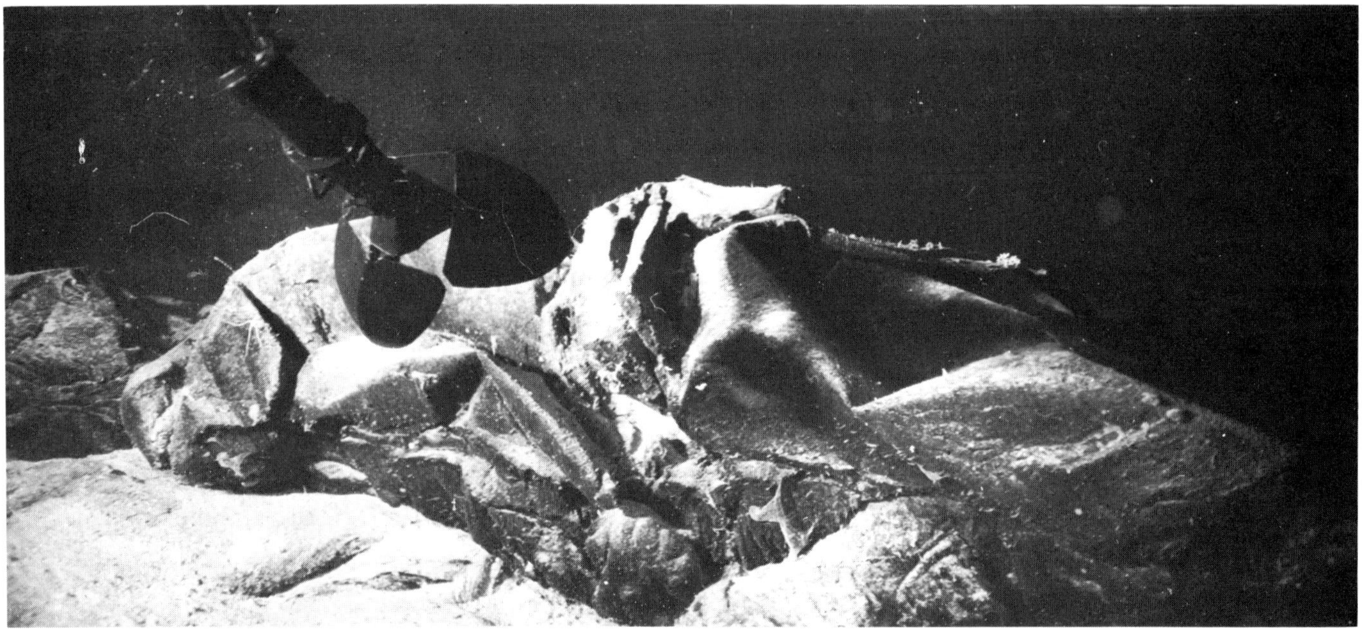

88 Wrinkled sheet flow. Photograph taken from French submersible CYANA on dive no. 74-30.

89 Apparent small sheet flow partly covered by sediment on steep slope.

Flow Foot Rubble and Talus

Flow foot rubble and talus is common at the base of the steep flow fronts of a pillowed flow (Figures 90–106). This rubble is so common at the foot of flow fronts that it is included here, despite the fact that it includes material of diverse origin and is, of course, not an individual cooling unit, but rather fragmental material derived from breakage of all types of cooling units mentioned previously.

The rubble is composed of extremely angular fragments of pillows ranging in size from greater than one meter to submillimeter-sized glass fragments. The typical fragment is shaped like a watermelon slice, preserving the outer curved surface of the pillow and bounded on the remaining sides by crude radial columnar joints formed by cooling-induced shrinkage. Other large fragments are the shelves of hollow layered pillows, as well as the fractured top crusts of hollow pillows. Many of the finer fragments are made up of the outer glassy rinds of pillows which spalled off during the growth of pillows (Tepley and Moore, 1974), during pillow implosions (Moore, 1975), or simply as a pillow fragment tumbled down the slope.

Much of the flow foot rubble undoubtedly formed during advance of the pillowed flow front. The steep slope of the flow front, and the fragile, fractured nature of the pillows draping the front, would cause many fragments to break off and tumble down the slope to accumulate as a talus pile at its base. Rubble

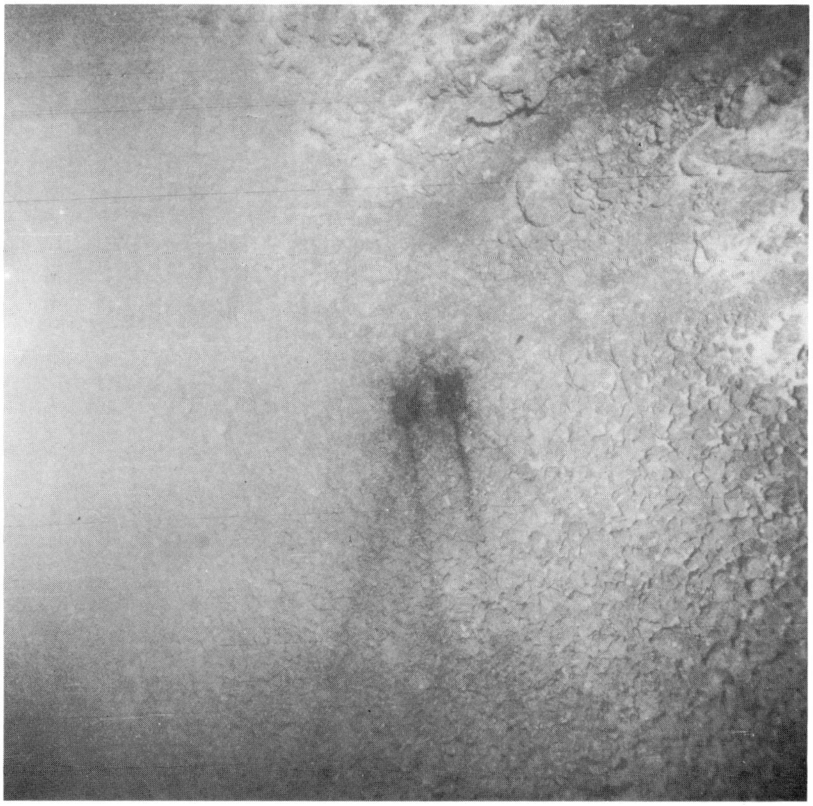

90 Flow foot rubble.

91 Vertical LIBEC photograph showing steep slope (sloping down to left) mantled with rubble and lava. Note in upper part of photograph, where lava is mantled with light-colored pelagic sediment, that dark trains of finer rubble extend downslope. Narrow dimension of photograph is about 22 m.

92 Vertical LIBEC photograph showing steep slope (sloping down to bottom) mantled with pillow joint-block rubble. Rubble blocks are covered by pockets of light colored sediment, and base of slope is largely covered by sediment. Narrow dimension of photograph is about 17 m mid-way downslope.

93 Pillow joint block rubble. Note fragment of hollow layered pillow in center foreground with two shelves.

of this kind would subsequently be overrun by the flow and must, therefore, be an important element at the base of a pillowed flow sequence.

In addition much of the rubble must have avalanched down the steep flow front as the result of later post-volcanic processes such as earthquakes, currents, biological action, sedimentation, and tectonic movement. This phase of rubble formation would produce material similar to the talus piles at the base of fault scarps, although it would have a more mono-lithologic character than that which is shed from fault scarps which cut several lava flows as well as rocks of other origin.

Most of the flow foot rubble is poorly sorted, although in some cases it appears to be composed primarily of block-sized fragments. This aspect is probably more apparent than real because the small-sized material has filtered down between the blocks and is not visible on the surface. In many places, however, sedimentation has filled in between the blocks and subsequent downslope movement and fragmentation have produced more fine material which is trapped on top of the sediment. Bottom currents and continued downslope movement have sorted this fine material further and produced small deposits of angular glassy sand and fine rubble.

94 Pile of relatively well-sorted, angular pillow fragments commonly bounded by radial cooling joints.

95 Pile of pillow joint block rubble. Gravel-sized pillow fragments are visible on right where they lie on top of sediment which prevents settling between larger blocks.

96 Fragments of elongate pillows which have slumped down to base of flow front.

97 Rubble of pillow fragments largely covered by sediment.

98 Gravel and block sized rubble. Photograph partly obscured by shadow of manipulator arm.

99 Well sorted pillow joint-block rubble. Fish on left is about 20 cm long.

100 Pillow joint-block rubble containing several tabular slabs which are fragments of shelves of hollow layered pillows.

101 Rubble of pillow fragments largely covered by sediment. Slabby fragment in upper right includes three shelves from hollow layered pillow.

102 Rubble slope almost entirely covered by faintly ripple-marked sediment. Gravel and sand-sized fragments of glassy pillow crust lie on sediment.

103 Steep rubble pile of pillow fragments. Gravel-sized glassy pillow crust fragments occur with sediment in sediment-filled chute.

104 Sand-sized pillow crust chips on sediment near margin of rubble pile.

105 Steep rubble pile with finer pillow fragments near margin of sediment-filled chute.

106 Block and gravel-sized rubble pile at base of scarp.

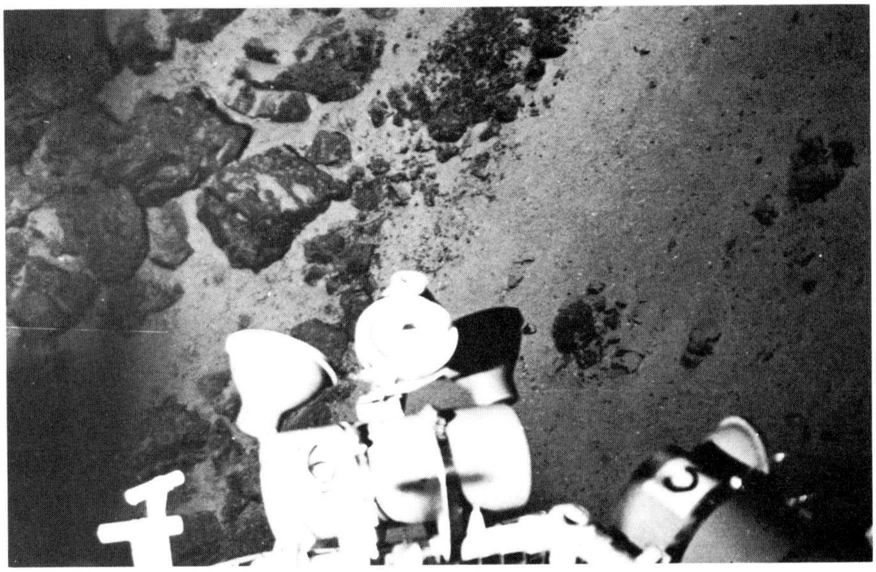

Features of Pillow Crust

The surface of fresh pillows are complexly marked by cracks, ridges, striations, and protrusions of various sizes and shapes (Figures 107–124). Observations of growing pillows in shallow water (Moore, 1975) demonstrate that many of these features are related to the growth of the glassy crust adjacent to spreading, accreting cracks (Figure 22). Such cracks form when the pressure of molten lava within the pillow breaks the outer brittle crust, and as the crack widens, and molten lava wells into it, cooling causes new crust to solidify on the trailing edges of crust.

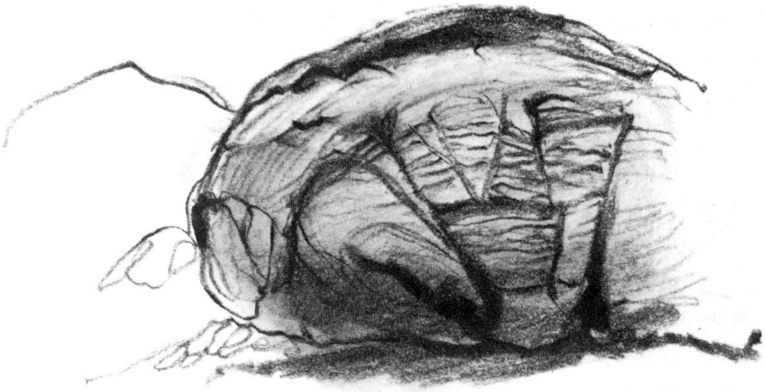

Spreading cracks apparently do not form in certain types of cooling units which are either very small or produced by very rapid delivery of lava, and consequently these units are notably smooth on their outer surface. The small pillow fingers of knobby pillows, generally 2–10 cm in diameter apparently grew by enlargement of a thin plastic skin and are quite smooth. Likewise sheet flows were fed rapidly, thinned out, and had little time to develop a brittle crust before flow ceased. They are generally smooth, but in some places the thin plastic crust became wrinkled in the

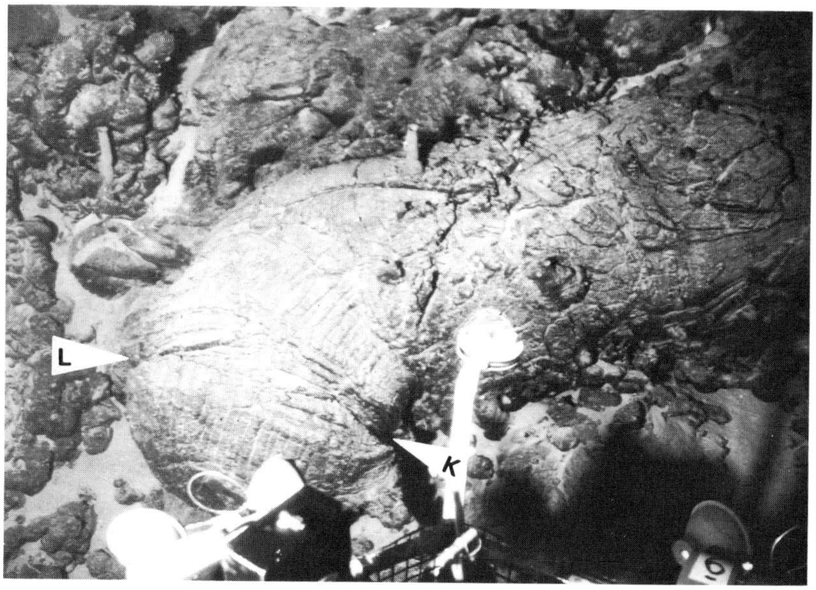

107 Corrugations and fault slivers on downhanging elongate pillow. Corrugations occur in matching areas of crust perpendicular to each of two spreading cracks. Fault slivers occur parallel to the crack on left and produce constriction of pillow.

108 Breadcrust incipient spreading cracks on bulbous pillow.

109 Spreading crack (L–K) in large pillow with clearly defined fault slivers parallel and close to crack, and more distant corrugations normal to crack. A second spreading crack parallel to axis of the pillow is visible on the right.

same manner as the ropy pahoehoe surface of terrestial lava flows (Moore *et al.*, 1971).

Generally, however, the outer crust of pillows grows to a thickness of about 1 cm or more during the period of growth of a flow lobe, and continued growth requires the brittle fracturing of the crust and the widening of these fractures. Hence, the roughness of the outer surface of the pillows is related to the thickness of the crust during growth. A corollary of this fact is that slow growth rates produce more ridged and fractured pillow forms.

Much of the pillow crust is generated by cooling and accreting adjacent to *spreading cracks.* They commonly have opened for a considerable distance, for

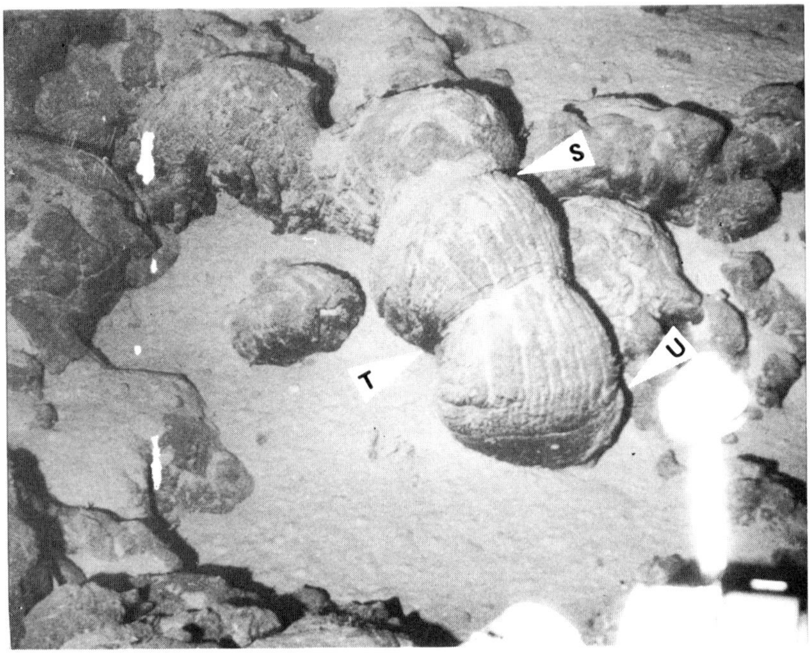

110 Two mirror-image pillow lobes which have grown by spreading on crack T. Before final stage of growth of this elongate pillow, fed from upper left, points S and U were adjacent and pillow was bounded by cap below U which was adjacent to S. Spreading on crack T generated the two nearly equal sized areas of new corrugated crust as crack T moved downslope at one-half the distance (and rate) of point U (or the cap of the pillow). Note that ridges and valleys of corrugations match on opposite sides of the spreading crack and that a few fault scarps parallel to the crack have developed. Note also that corrugations become less prominent and divide into two or more ridges with distance from the spreading crack T.

111 Spreading crack AD on bulbous pillow. Note matching corrugations formed at crack on each side normal to crack. Well-developed fault slivers form parallel and close to crack during final stages of spreading before crack freezes shut. At this stage crust in close to crack is thicker because of very slow spreading and fails by brittle fracture instead of plastic bending.

example 20 cm, and hence have formed new crust on each side 10 cm wide. Such cracks generate 2 types of surface ridges (Figures 22 and 107) clearly visible in photographs: *corrugations* and *fault slivers* (Moore, 1975). Corrugations are formed where opening on the spreading crack is moderately rapid (approximately 5 cm/sec) and the ridges form at right angles to the spreading crack. The corrugations are generally 1–10 cm in wave length and 1–3 cm high. Where spreading is symmetrical, corrugations are formed on both sides of the crack and distinctive ridges and valleys can be matched on opposite sides of the crack.

Observations in shallow water (Tepley and Moore, 1974) indicate that fast spreading cracks (approximately 5 cm/sec) are zig-zag and irregular in plan and this irregularity may be important in the generation of the corrugations. Perhaps delivery of molten lava is higher in wider parts of the crack and ridges are formed opposite them.

Generally a single corrugation is quite uniform outward from the spreading crack and can be traced for a distance 10–20 times its width. However, in some cases the corrugation ridges become less prominent and divide into two or more smaller ridges outward from the spreading crack. If we assume that spreading on the crack slows down before stopping, then the larger corruga-

112 and 113 Two views of the same bulbous pillow with clear remnant of spreading crack (L). Crust generated at this crack has a grid-like pattern because of intersection of corrugations normal to crack and faults parallel to crack (L). Knobby pillows visible in background.

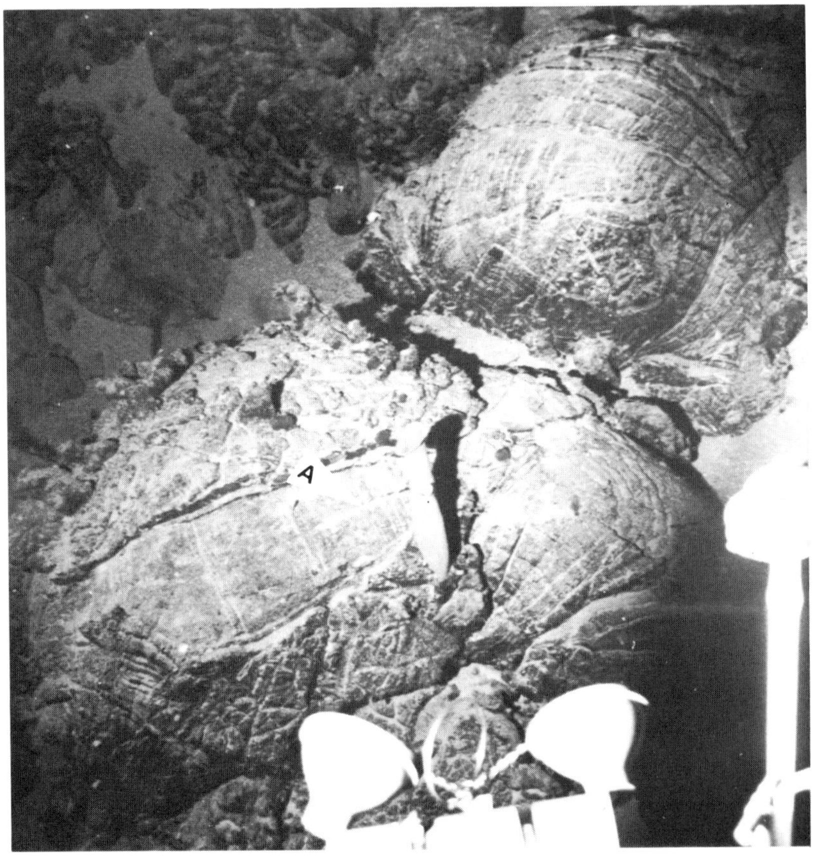

114 Double pillow (center) with well-preserved fault slivers which parallel spreading crack at the plane of pillow symmetry. Note that pillow is markedly constricted where fault slivers are well developed adjacent to spreading crack. Photograph taken from French submersible ARCHIMEDE.

115 Bulbous pillow with clearly defined younger patch of crust which has grown by spreading on a crack which has left a faint scar (A). Faint, matching corrugations are perpendicular to crack (A).

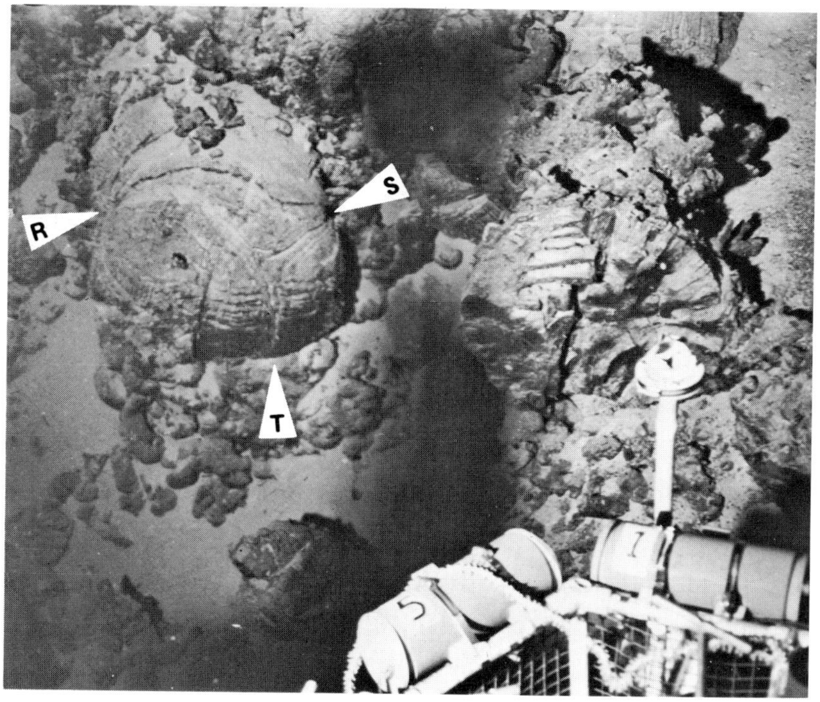

116 Pillow in which three intersecting spreading cracks (R, S, T) have opened and generated new crust nearly doubling previous surface area of pillow. In center of photograph (near compass) is exposed a broken hollow, layered pillow with 5 subhorizontal shelves. Note that the thickness of the side wall increases downward in this hollow pillow.

tions would be favored by somewhat slower spreading. This is supported by the fact that rapid flow tends to produce a noncorrugated surface as in the case of a sheet flow fed when the confining walls of a hollow pillow fail.

Fault slivers form parallel to slow opening cracks (approximately 0.2 cm/sec). They are generally less than 1 cm wide and are bounded by small normal faults of a few millimeters displacement, down-thrown on the crack side. They are symmetrically arranged on each side of the crack and comprise small fault blocks tilted away from the crack. The innermost faults encompass a small graben at the site of the crack. The faults steepen with depth and commonly become overhanging; that is, the normal faults pass into reverse faults with depth. The faults were apparently produced as molten lava within the pillow rise in the crack area and dragged up the solid crust a short distance from the crack axis. The thinnest, youngest crust adjacent to the crack was not uplifted because rising material was diverted laterally to accrete to the diverging crustal plates.

Some pillows grow at a rate intermediate between that of corrugation and fault development and both structures are formed and intersect at about right angles giving an overall squared pattern to a pillow surface. In other cases, widespread fault scarps occur in pairs opposite a spreading crack on otherwise smooth pillow crust. They seem to mark hesitations during relative rapid spreading.

The fine texture of the pillow surface is in some places quite smooth and flat adjacent to spreading cracks on which spreading has been rapid enough to pro-

117 Phallic pillow which (despite spalling of glassy crust) preserves scars of several spreading cracks with corrugations perpendicular to them. Last spreading crack is probably the one on blunt end of pillow.

118 Large pillow (behind compass) with prominent spreading crack. Note that broad corrugations perpendicular to crack (vertical in photograph) match on each side of crack.

119 Complex intergrown pillows with several spreading cracks. Crack DE produced a narrow segment of crust at the pillow base and perhaps much of the basal crust. Crack FG spread asymmetrically producing a small segment of crust below it and much of the visible crust above it.

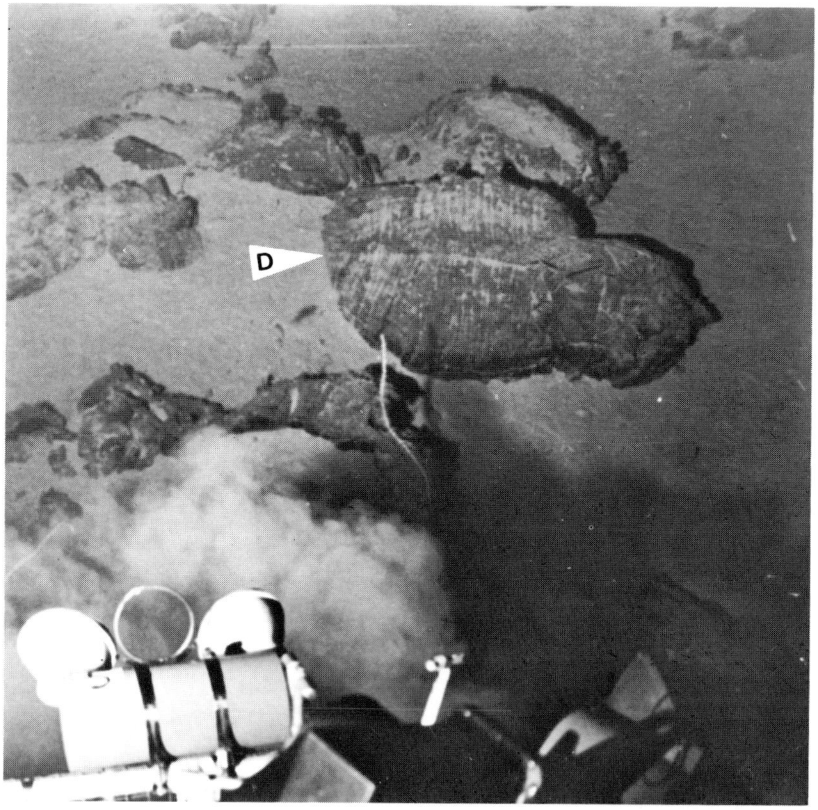

duce no corrugations or only very faint ones. This crust when examined closely shows a minutely lineated texture perpendicular to the spreading crack. Microscopic study of pillow surfaces shows that this lineation is produced by small drawn-out vesicles and filaments of glass generally smaller than 1 mm in width. The vesicles and glass filaments show a shingled aspect, dipping into the surface at a small angle. The dip of these features is *away* from the spreading crack.

The overall aspect of pillow shape shows a general relation to the character of its crust. Where the surface is corrugated, it is generally inflated and convex. Where the surface is cut by fault slivers, it is generally constricted and concave. This difference results from speed of growth and consequently the greater degree of plastic inflation of thinner, corrugated crust. These differences partly explain the knob-and-neck shape of elongate pillows (Figure 22). Commonly the knobs are corrugated and the necks faulted.

The final growth of some pillows appears to take place on a system of *breadcrust cracks* (Figure 108) that have only opened a small amount (less than a few cm). Such cracks commonly intersect one another and radiate from centers of greatest expansion. They commonly have matching walls and the broken plates of crust can be visually moved back to their original positions of juxtaposition. These cracks are shallow (approximately 1 cm) and rehealed by fresh lava which has welled into the crack from below the fractured plates of crust.

120 Large pillow in which most of visible crust was produced by spreading on crack (D).

121 Knobby pillow which has expanded after initial growth by spreading on two cracks (O and P).

122 Trapdoor pillow with several sediment-filled interconnected cracks.

123 Pillow with incipient spreading cracks.

124 Knobby pillows with breadcrust cracks on surface crust.

Faults and Related Tectonic Features

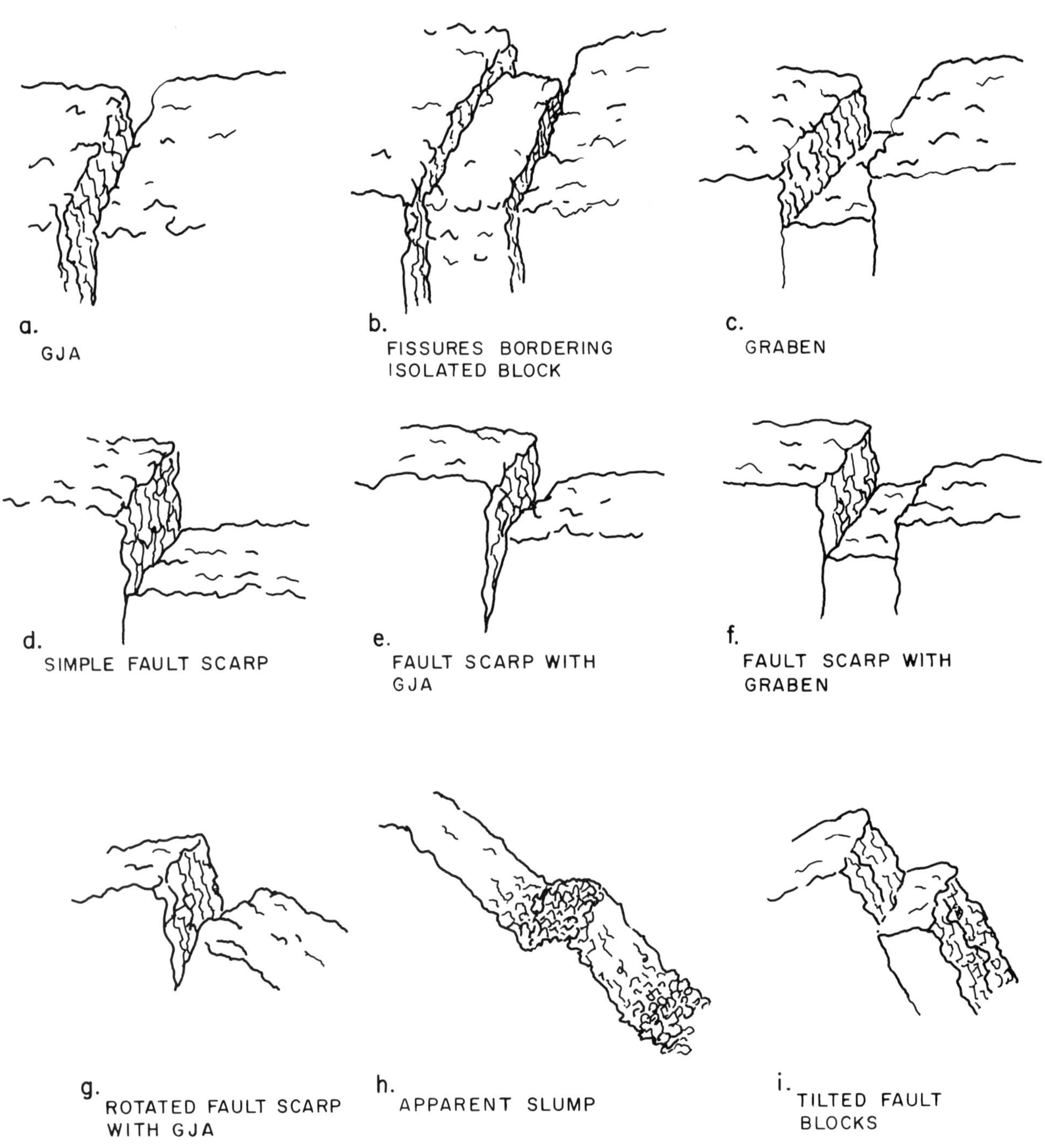

125 Principal tectonic features observed in the study area. a–c are pure tensional features with vertical fault planes. These features occur throughout the inner floor, but are the dominant tectonic elements in the central zone of volcanic activity. d–g are examples of tensional features with dip-slip displacement and are the dominant tectonic elements in the marginal tectonic province flanking the central volcanic zone. h and i are examples of the fault scarps observed on the east and west valley walls. (After Ballard and van Andel, 1977.)

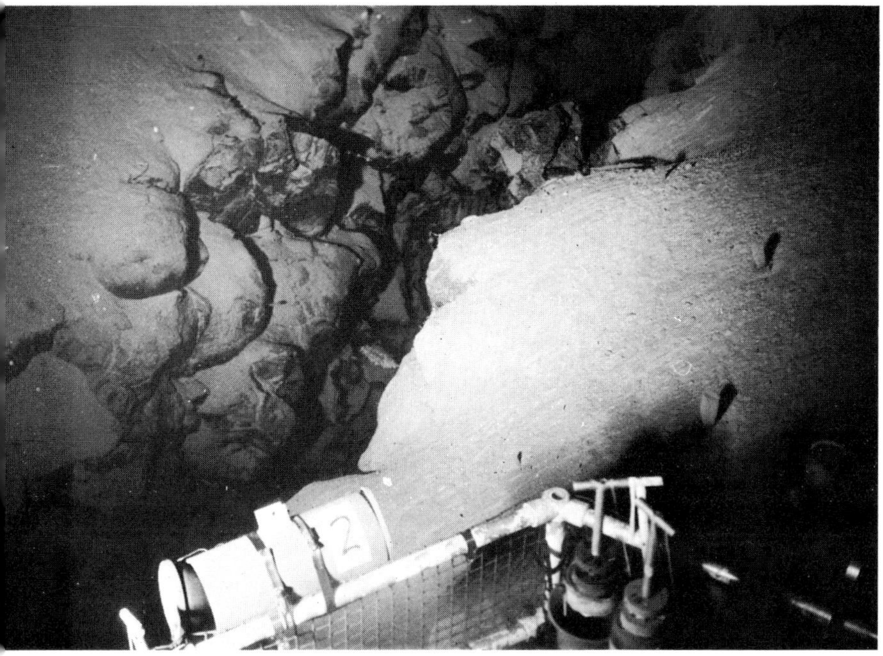

126 Simple tensional fissure of the inner rift valley floor, the most common feature encountered in the study area.

127 Artist concept of a gja (Figure 125a). The fault plane is vertical and exposes numerous pillow lava cross sections having a radial joint pattern.

The newly formed oceanic lithosphere is continuously transported laterally away from the central zone of intrusion and volcanism. As the lithosphere cools and hardens, it is subjected to the tensional forces of plate separation. Fracture planes develop in the brittle rock, dividing it into a series of fault bound blocks. These blocks then undergo horizontal movement which displaces them east and west away from the active zone of injection. The fractured blocks pass through the tectonic provinces flanking the center zone to the base of the inner rift valley walls. Here they undergo vertical uplift and are incorporated within the flanking crestal mountains. Vertical uplift appears to occur in incremental stages as the position of the east and west walls jump inward with the elevation of a portion of the floor (Ballard and van Andel, 1977). These fault bound blocks consist of individual volcanoes which formed along the central axis of injection. As the lithosphere moves outward, the vertical component of transport is reversed and the blocks move downward, eventually reaching the abyssal depths of deep ocean basins. During this process of transport, the upper crust undergoes mechanical and chemical change which alters its original composition and physical characteristics. These changes, including the nature of crustal fracturing, the degree of hydrothermal alteration, and the effects of subsequent sediment burial are the subject of many investigations. A critical factor in these studies, however, is a fundamental knowledge of the crust's initial physical starting condition. How, for example, is the crust initially fractured and what are the physical dimensions of the crustal blocks which leave the inner rift valley?

Several investigations of the results of Project FAMOUS have examined the tectonics of the rift valley (Arcyana, 1975; Ballard and van Andel, 1977; Ballard et al., 1975; Luyendyk and Macdonald, 1977; Macdonald, 1977;

Macdonald and Luyendyk, 1977; Needham and Francheteau, 1974; Ramberg and van Andel, 1977; Ramberg, Gray, and Reynolds, 1977). However, none of these studies contain a detailed description of the various tectonic elements into which the floor of the inner rift can be divided. Ballard and van Andel (1977) outline the small scale tectonic elements (Figure 125) but due to the limitation of space proper documentation was impossible. In the following pages, therefore, these individual features are described by employing a series of photographs and illustrations of the dominant features observed on the rift valley floor and walls.

128 A small fissure running across an otherwise undisturbed volcanic terrain. A pillow in the center of the photograph is cut by the fault plane and suggests that vertical displacement of a few centimeters has occurred with the foreground having dropped down relative to the background.

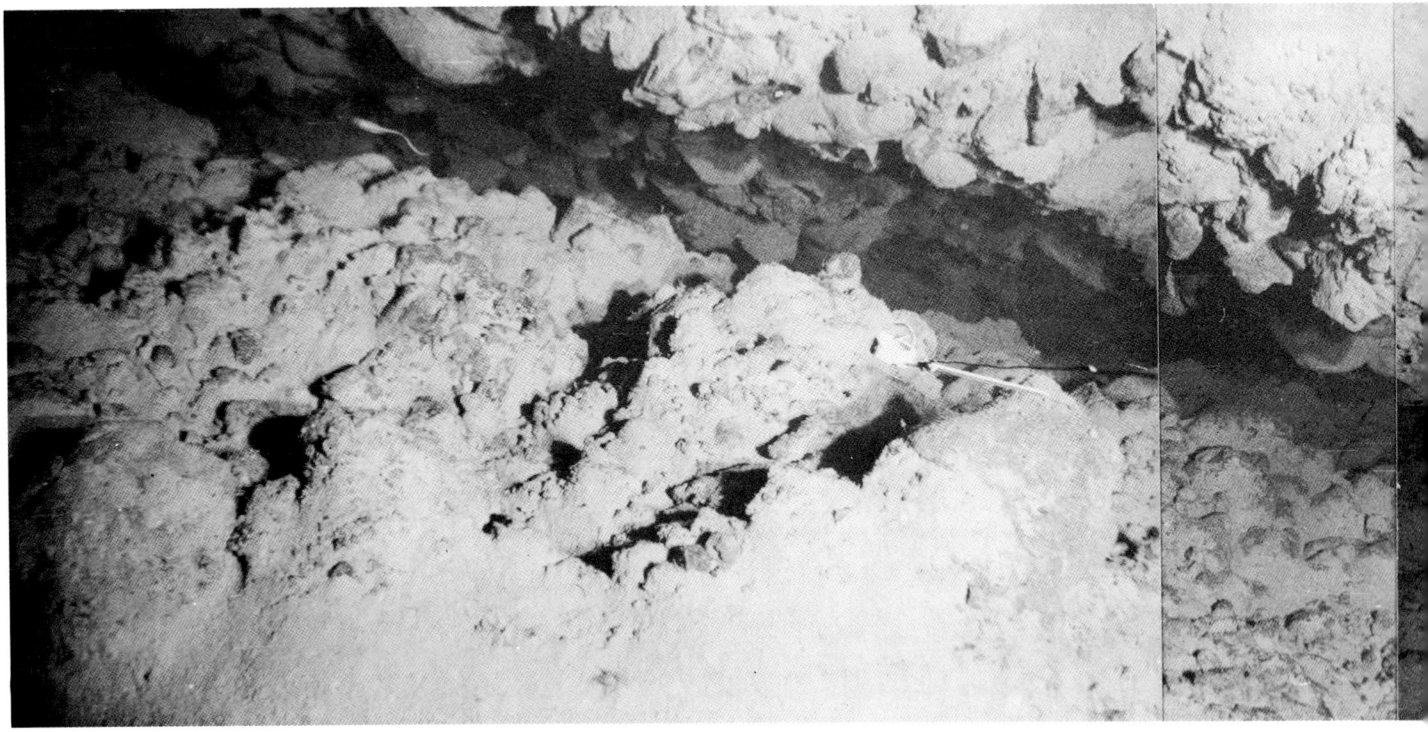

129 Another example of a small fissure cutting across individual pillow features.

130 A mosaic constructed from pictures taken with a remote camera lowered from the research vessel KNORR showing gja. The man-made object seen three times in the photograph is a compass which is suspended beneath the camera. Unfortunately, the compass was damaged earlier in the tow and is not functional. Note that the fissure does not strike in a straight line but has a curved form. The photographs were taken in a similar fashion to those obtained with the LIBEC camera only they cover a smaller area.

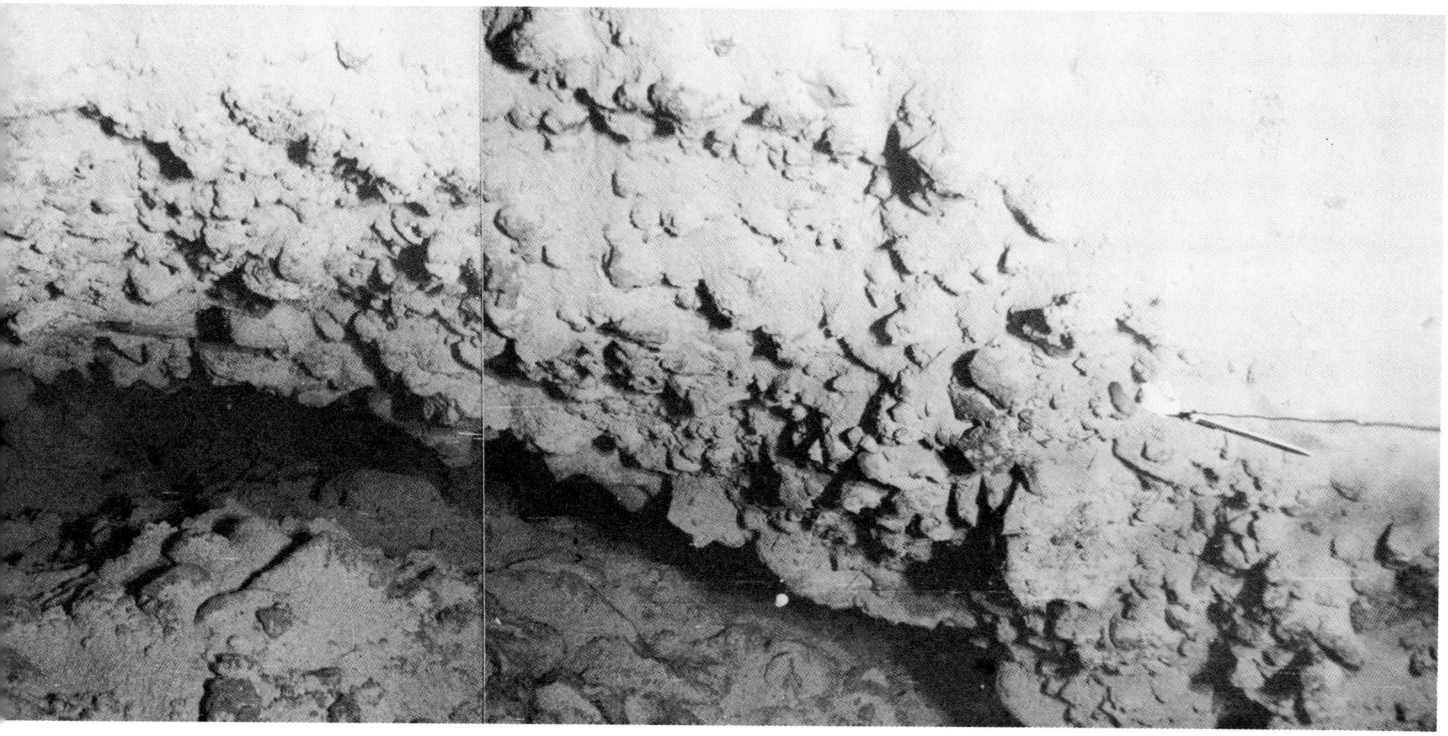

Tensional Fissures

Gjár

The most common form of tectonic rupture in the central volcanic province are fissures or *gjár*. They have undergone horizontal separation, but exhibit no vertical displacement (Figures 126 and 127). The strike of these features is primarily parallel to the valley axis although other related trends occur. A complete analysis of the trends (Ballard and van Andel, 1977) shows that all the fissures (as well as faults) mapped by submersible observation can be attributed to a single stress field.

The width of the gjár vary from a few centimeters to greater than four meters. Their width, however, shows no relationship to their location within the central volcanic province or the marginal tectonic provinces to the east and west where gjár also are common. The fault planes generally are nearly vertical and cut across individual volcanic forms (Figures 128 and 129). Frequently it is possible to see parts of the same truncated volcanic form on either side of the gjár. In the numerous instances that such correlations were made, no translational motion was observed (Figures 130, 131, and 132). In many cases, a gja has opened 3–4 meters but delicate pillow forms still remain perched out over the edge of the fissure (Figure 133). The presence of these features which are brittle and easily broken suggests that fissure development can take place rather gently without pronounced shaking of the bedrock.

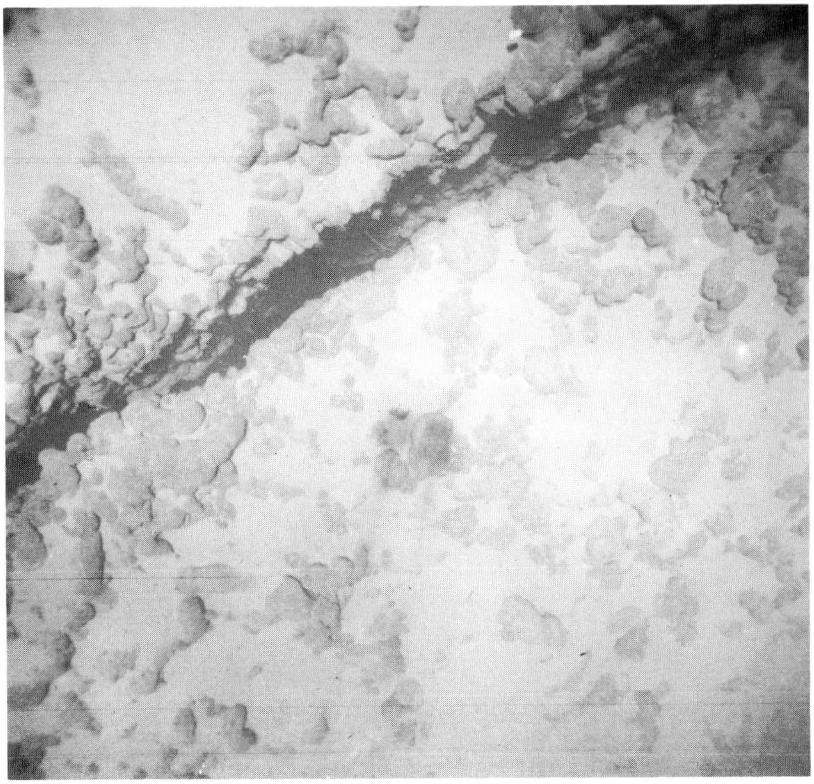

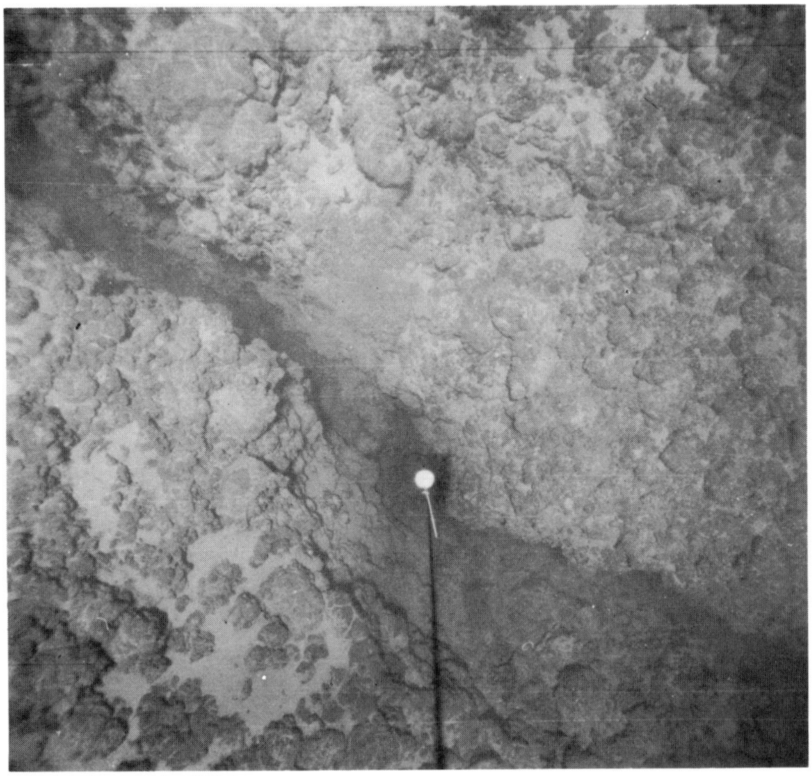

131 LIBEC photograph of a gja. Note the undisturbed character of the surrounding volcanic terrain.

132 LIBEC photograph of a gja.

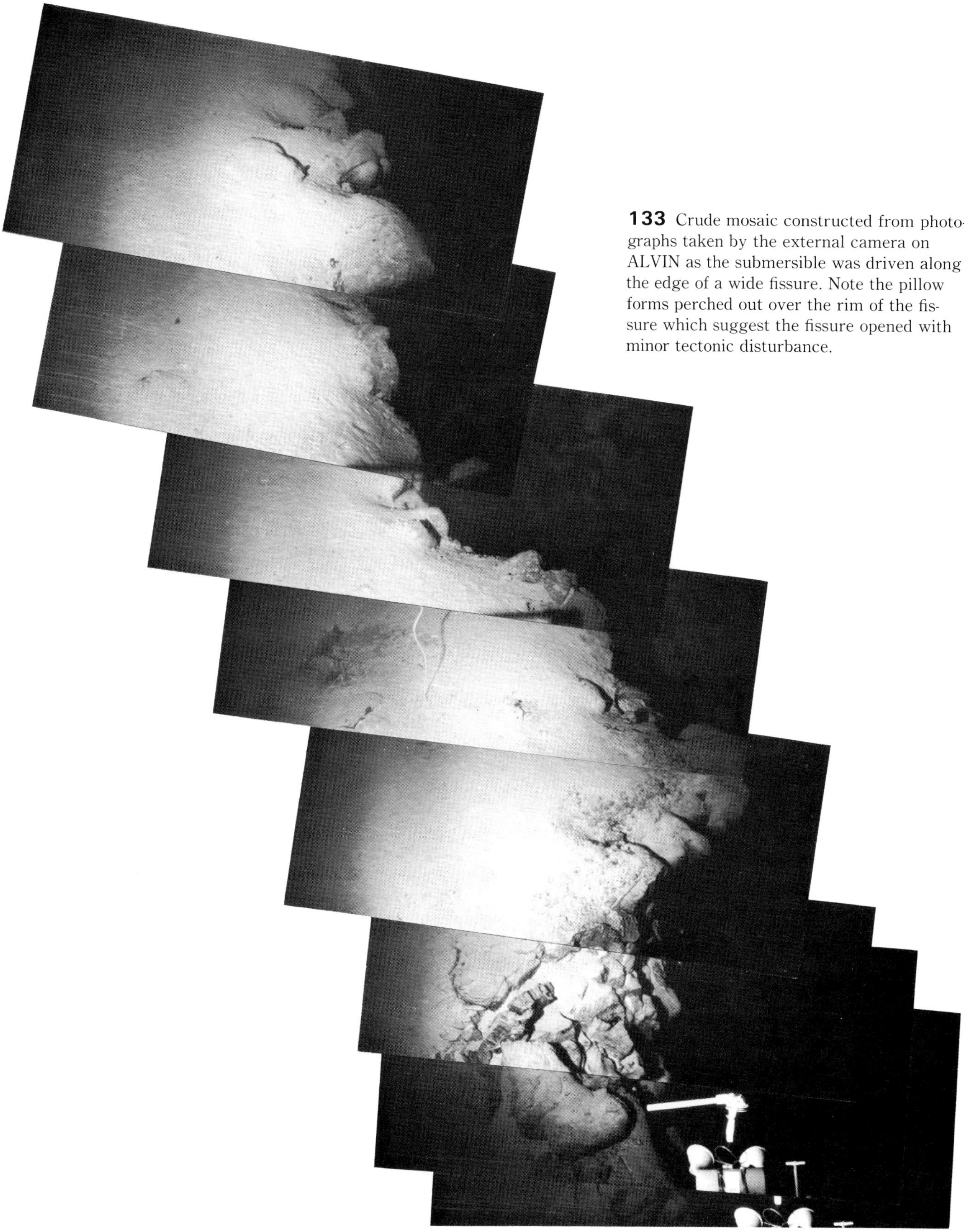

133 Crude mosaic constructed from photographs taken by the external camera on ALVIN as the submersible was driven along the edge of a wide fissure. Note the pillow forms perched out over the rim of the fissure which suggest the fissure opened with minor tectonic disturbance.

Isolated Fault Block

Observations from ALVIN as it drove along the strike of many of the gjár revealed considerable variation in width and degree of fracturing (Figures 134 and 135). A single fissure, for example, would divide and isolate a small down-dropped block of volcanic terrain to form an *isolated fault block* (Figure 136). Where the fissure divided, secondary fracturing is intense and produces a widening of the fissure opening. If only a portion of the isolated fault block is seen in a bottom photograph, the second branching fissure can be mistaken for a different stress plane instead of simply the boundaries of a fractured block (Figure 137). In the submersible, however, it is possible to drive along the feature and determine its field relationship; eliminating such errors (Figure 138). Isolated fault blocks graded upward in size to large portions of the floor which were fault bound.

134 and 135 Two photographs taken a short distance apart of the same fissure illustrating the variation in width which frequently occurs along open fissures.

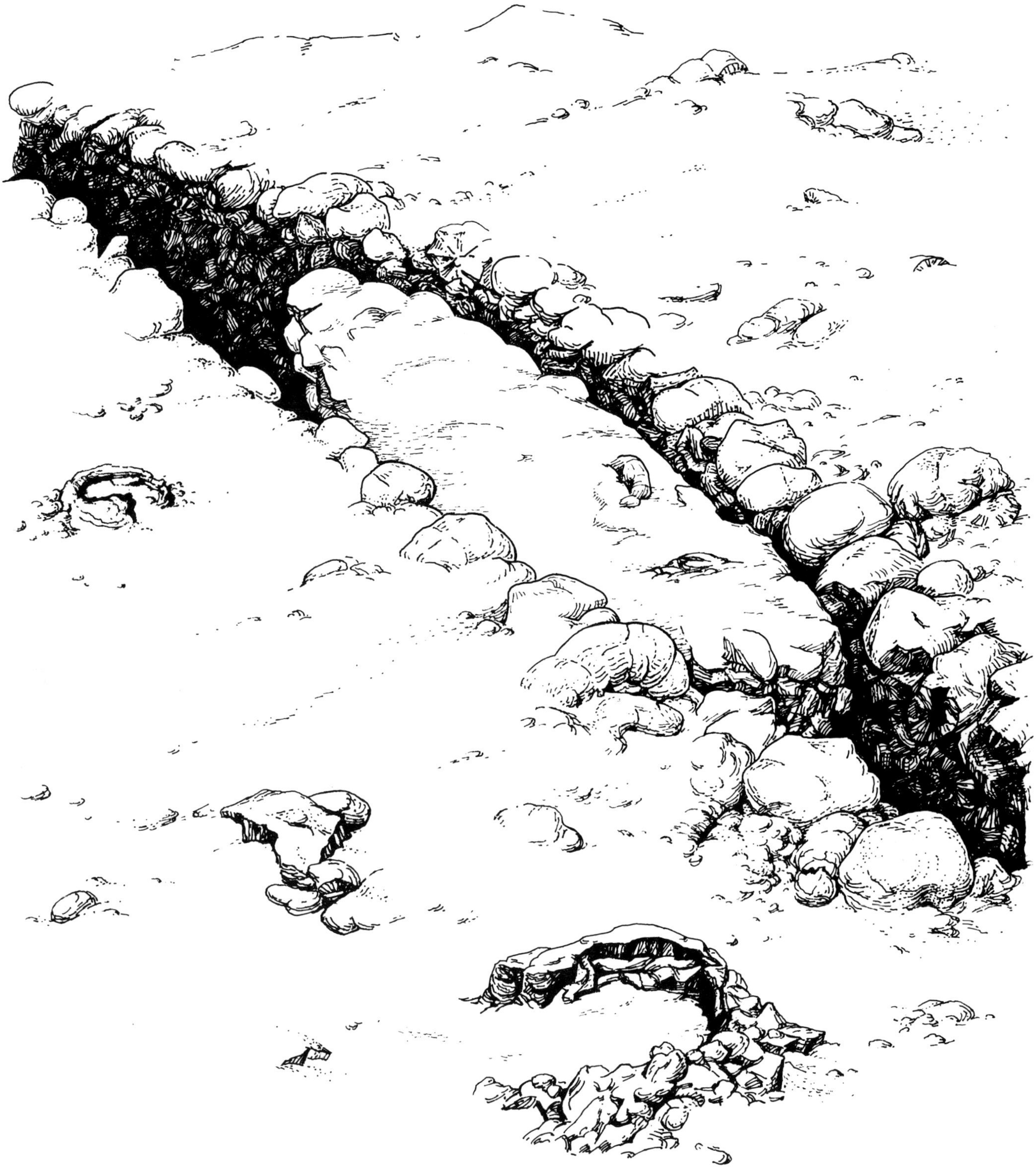

136 Artist concept of an isolated fault block within a tensional fissure (Figure 125b). The widest portion of the fissure occurs at either end of the enclosed block where brecciation was most severe.

137 A portion of an isolated block can be seen in the lower left hand corner of this LIBEC photograph.

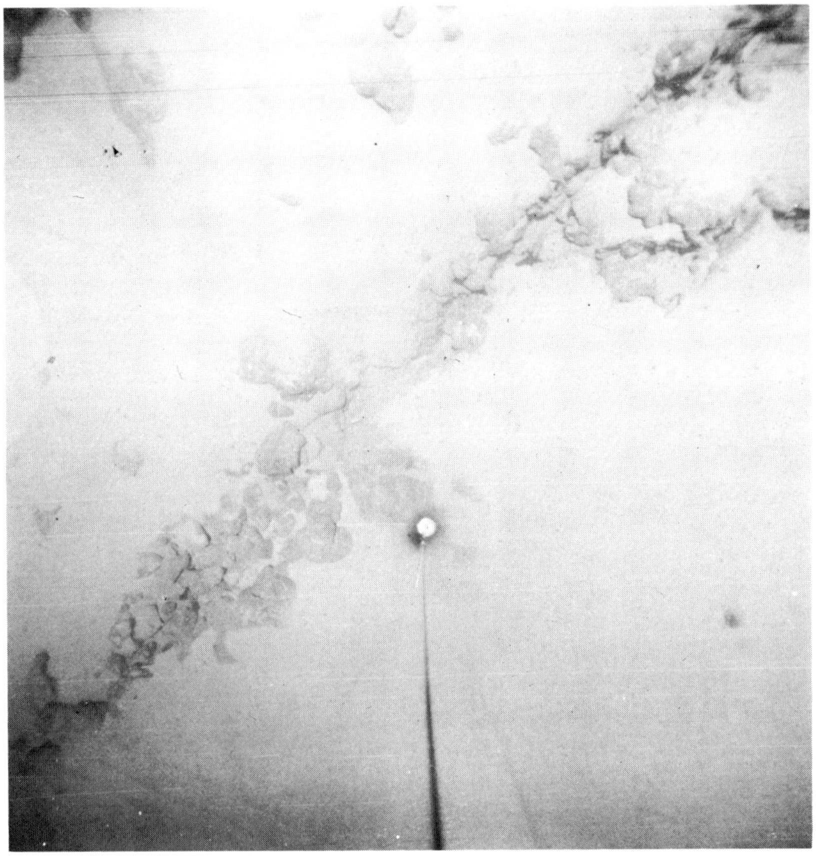

138 While driving along the strike of a fissure, the fault plane was observed to bifurcate for a short distance, isolating a faulted block. One fault runs down the center of the photograph while the second is in the upper right-hand corner.

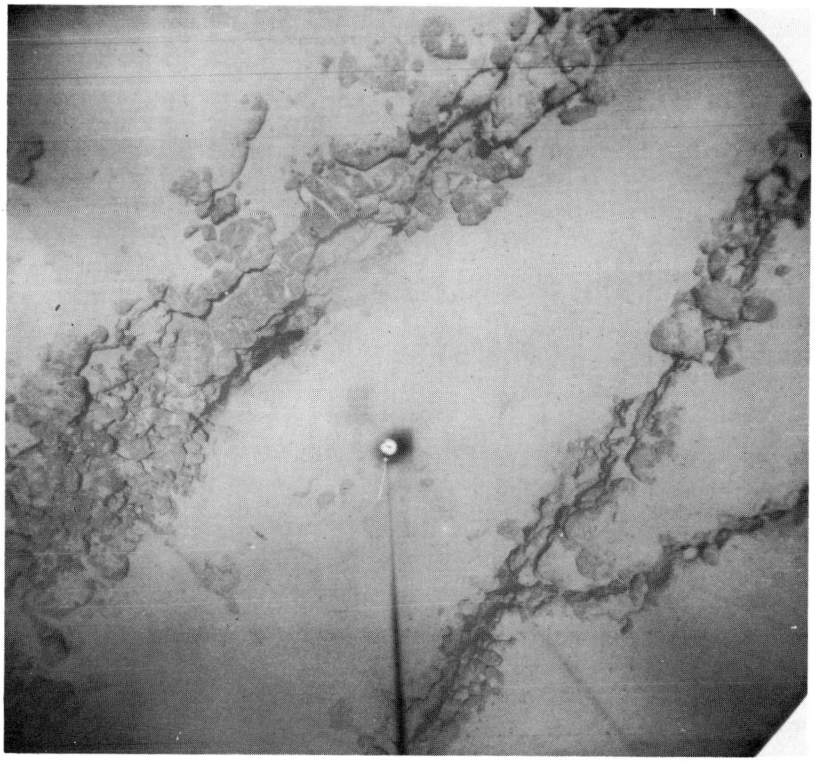

Small-scale Grabens

In many cases, cracking or fissuring was followed by collapse producing fault-bounded *small-scale grabens* (Figure 139). The floor of the graben was frequently covered with talus and sediment, but a surprisingly large percentage are flat-floored where the block had simply dropped down with little or no secondary fracturing (Fig. 140).

139 An artist concept of a small-scale graben (Figure 125c). In many instances, the surface of the lower block was similar to that of the surrounding volcanic terrain while in other cases it was covered with talus fragments originating from vertical motion along the bordering faults.

140 A look directly down at a small-scale graben provided by this LIBEC photograph. Overlapping photos on both sides also show the "graben" to be tilted.

Fault Scarps of Inner Floor

Fault scarps of the inner rift valley between the inward facing east and west walls are common. The amount of vertical displacement along the fracture plane increases away from the central axis. The dip of the fault plane also decreases from 90° in the central zone to about 60° at the base of the bordering walls. Since the steepness of the scarps is similar to that of many lava flow fronts, it is important to develop criteria for recognizing the differences between faults and flow fronts. Since the flow fronts are constructive volcanic features, their most important characteristic is downward sloping elongate pillows which are coherent in form (Figure 141). Fault scarps on the other hand exhibit a distinctively different exposed rock surface. Since the fault plane cuts across the pillow forms, the scarp contains numerous pillow cross sections which have a characteristic radial joint pattern related to the cooling history of the molten lava (Figure 142). The number of pillow cross sections appears to reach its maximum when the strike of the fault is perpendicular to the lava flow direction and the least when the fault plane is parallel to the flow. As a result, there is a wide variation in the appearance of the fault scarps ranging from a high percentage of truncated pillows (Figure 143) to a high percentage of coherent pillow forms (Figure 144). When coherent pillows dominate, it can be difficult to determine whether the slope is fault-controlled or is a flow front. In some instances, it appears both processes have taken place (Figure 145).

141 A typical view of a lava flow front surface containing a mixture of bulbous and elongate pillow forms.

142 A typical view of a vertical fault scarp containing numerous pillow lava cross sections having a radial joint pattern.

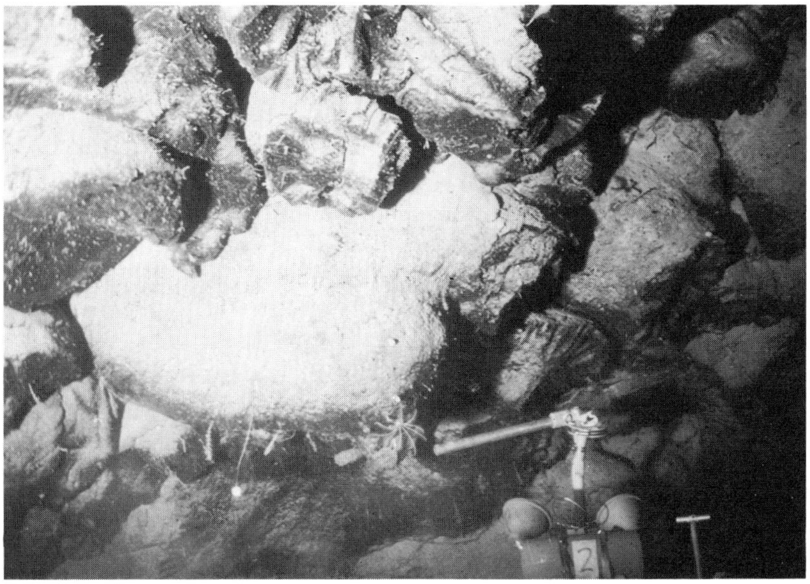

143 Vertical fault scarp where the number of truncated pillows exceeds the number of coherent forms.

144 Vertical fault scarp where the number of coherent forms exceeds the number of truncated pillows. It is in these cases that differentiating lava flow fronts from vertical fault scarps becomes difficult.

145 The western flank of Mt. Pluto which appears to have originated as a result of both volcanic construction and tectonic uplift.

146a and b Two artist concepts of a simple fault scarp commonly encountered in the inner floor (Figure 125d). Note the truncated pillow cross sections and the absence of talus at the base of the scarp.

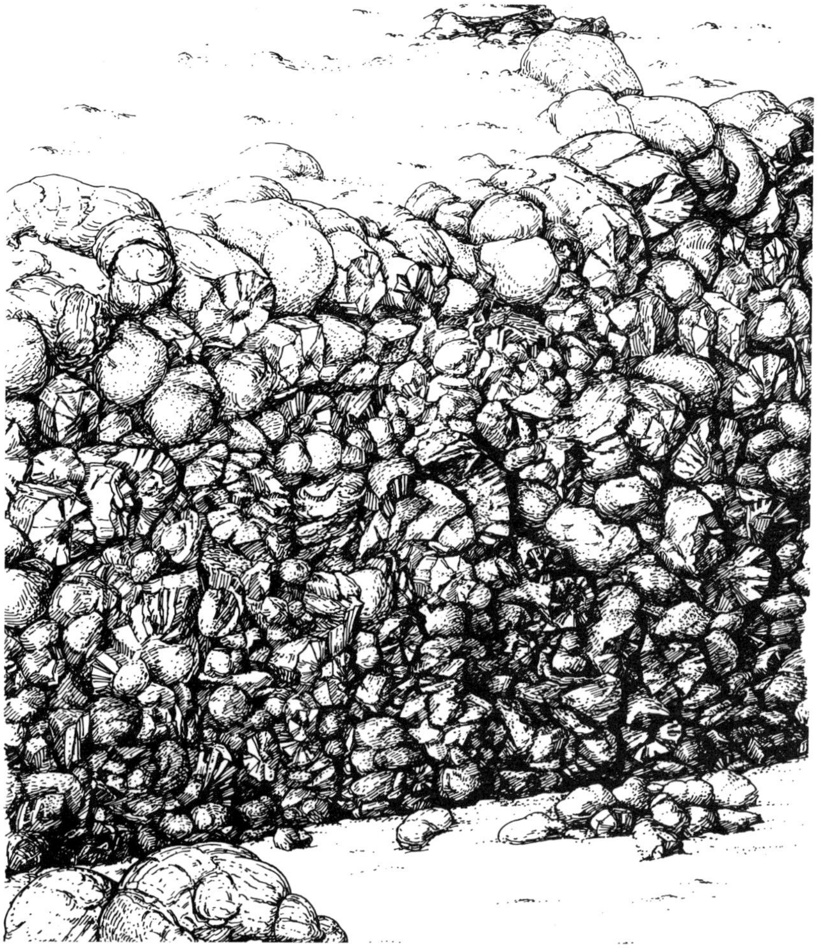

Simple Fault Scarps

The most common scarps in the inner floor are *simple fault scarps* which vary in height from less than a meter to over 40 meters (Figures 146 and 147). Except for a few small scarps cutting lithified sediment, the rock unit exposed in all these scarps is a vertical section of truncated pillow lava (Figure 148). No deeper bedrock layers were seen within the inner valley nor were any layers of interbedded sedimentary material observed in the more than 50 sections inspected. This absence of interbedded material in regions where pockets of surface sediments are abundant suggests that the pillow section is built over a relatively short period of time. Most of the small scarps had little to no talus accumulated at their base, implying that uplift can take place without major dislocation or slumping (Figure 149).

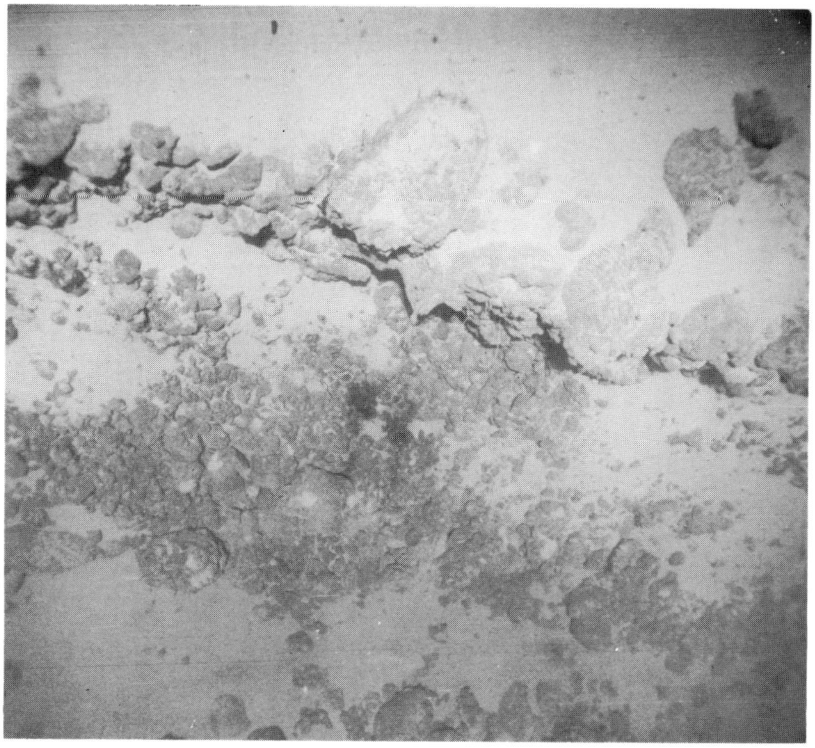

147 Simple fault scarp viewed from directly overhead by the LIBEC system.

148 View of vertical fault scarp. No interbedded sedimentary material was ever observed in any of the numerous scarps inspected, although they were commonly dusted with recent sediments which also accumulated at their bases.

149 The absence of talus accumulations at the base of these fault scarps suggests that vertical uplift has occurred without major violent activity.

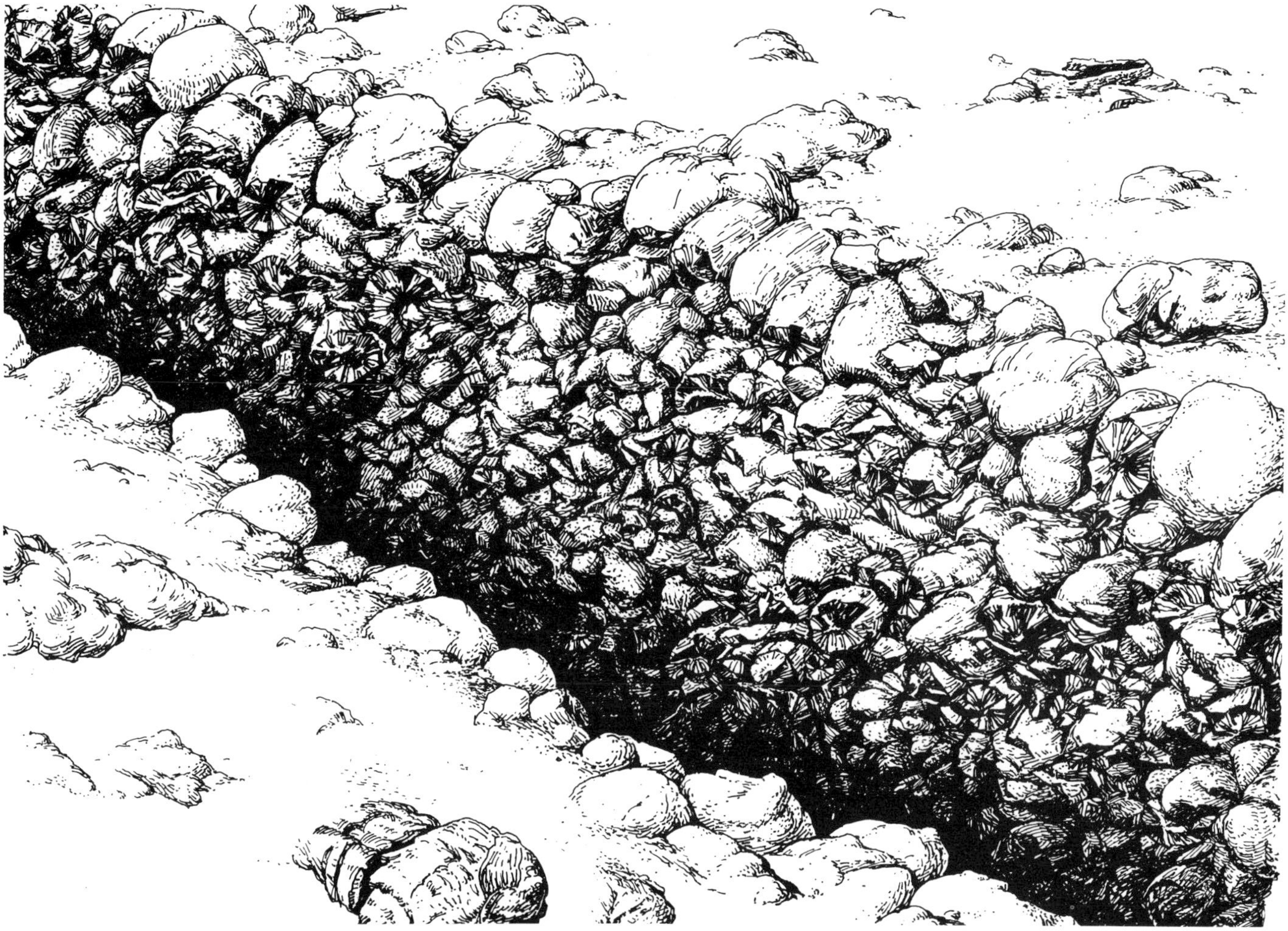

150 An artist concept of a fault scarp with a fissure at its base (Figure 125e). This tectonic feature occurs throughout the inner floor, but is dominant in the region between the central volcanic zone and the valley walls.

151 Example of a small fault scarp with an open fissure at its base.

Fault-gjár, Fault-graben Combinations
In addition to exhibiting vertical displacement, many of the scarps also reflect tensional stress. The result is the formation of *fault-gjár, fault-graben combinations.* A large percentage or the scarps in the marginal tectonic province between the volcanic axis and the inner valley walls have fissures at their base (Figures 150 and 151). Since talus formation appears minor on the smaller scarps, these fissures are not filled with rubble (Figures 152 and 153). Grabens which also form at the base of fault scarps average 2–4 meters in width and depth (Figure 154). Open fissures frequently border the down-dropped block which either have a coherent lava surface or are covered with talus and sediments (Figure 155).

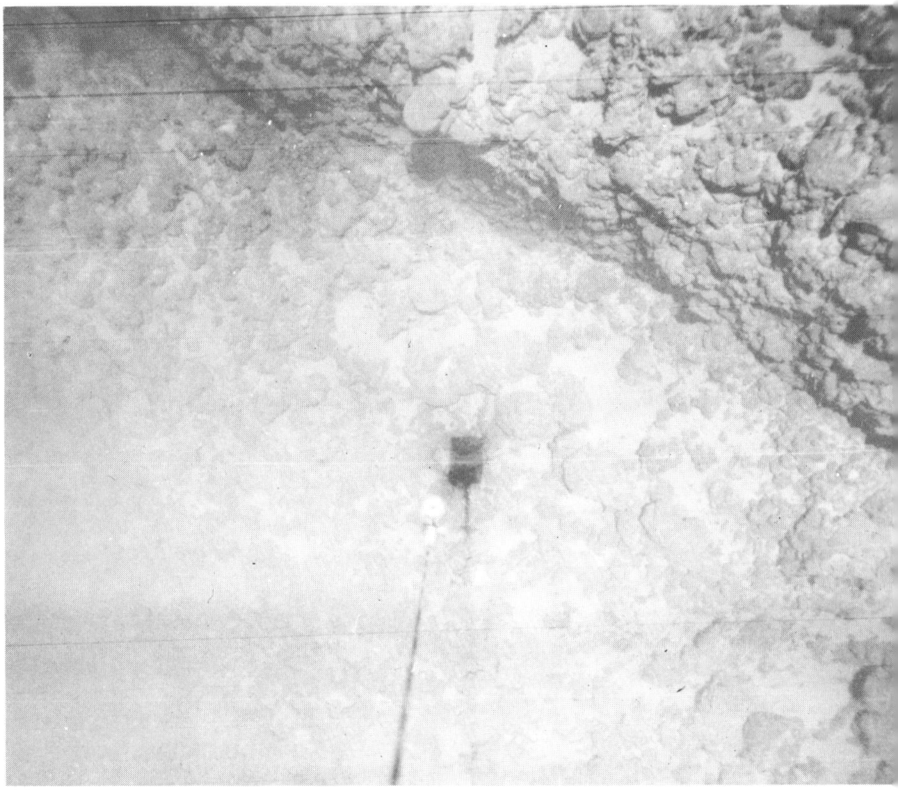

152 A view from LIBEC of a fissure-fault combination on the side of a volcanic slope.

153 The fissures at the base of fault scarps are obvious traps for sediment and talus. Few, however, appeared filled suggesting either gentle vertical uplift or the open fissures extend to considerable depth and require a large amount of material to fill them.

154 An artist concept of a small graben at the base of a fault scarp (Figure 125f). These graben commonly have a flat sediment floor while in other instances are partially filled with talus rubble. In this drawing the dark streak corresponds to an accumulation of pteropod shells which were often observed.

155 LIBEC photograph of a small graben at the base of a small fault scarp. Note the absence of talus.

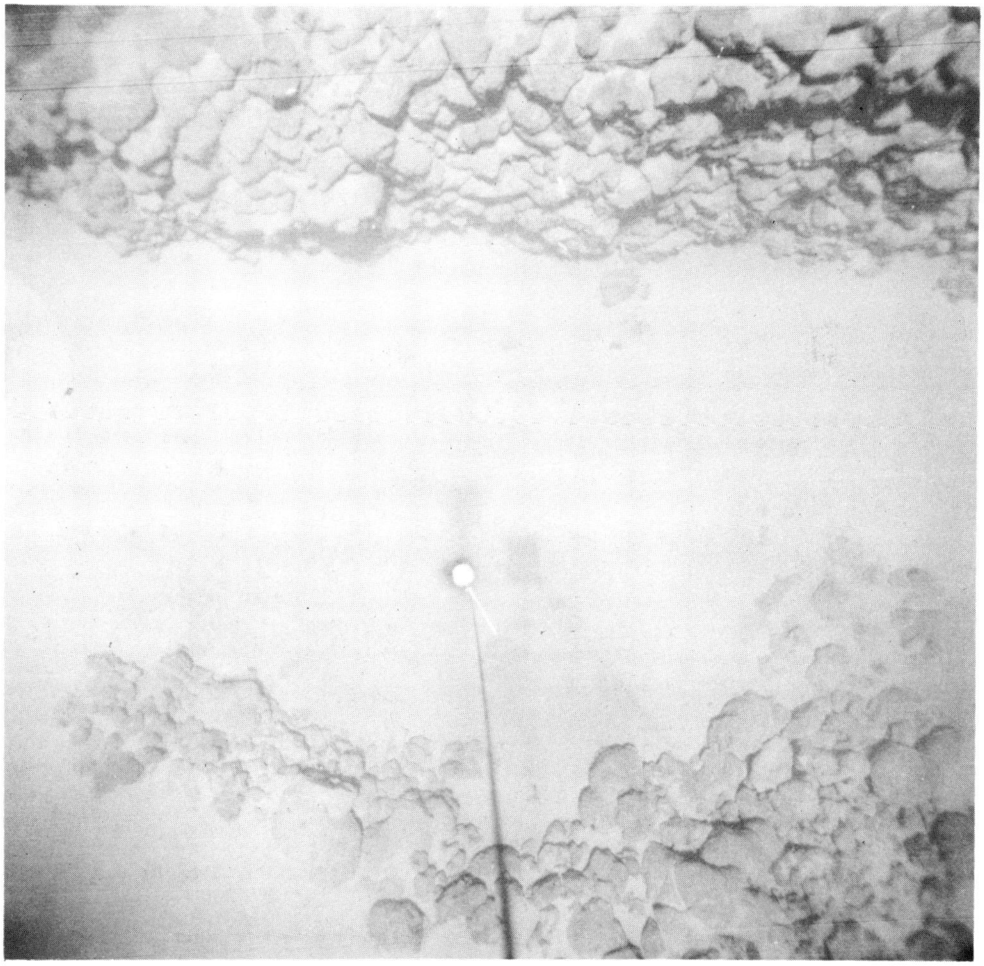

Rotated and tilted blocks

In addition to horizontal and vertical motions, some of the faults exhibit varying degrees of rotation and tilting. *Rotated blocks* rarely occur, and the few instances documented were associated with regions of severe faulting (Figure 156). In such cases, many of the smaller blocks were observed to have rotated or pivoted away from a larger bedrock unit (Figure 157) which in extreme examples also produces overhanging or back tilted fault planes. Rotated blocks are rotated around an axis perpendicular to the fault plane. *Tilted blocks* are by far more common (Figure 158) with their frequency of occurrence increasing towards the base of the east and west walls. Tilted blocks are rotated around a near horizontal axis parallel to the fault plane. Where the degree of tilting was measured, it was found to range between 5 and 7 degrees.

156 An artist concept of a rotated fault scarp with associated open fissure. The rotation was primarily away from the fissure fault plane although one major scissor fault was observed in the region of Mt. Uranus (Figure 7).

157 An example of a rotated fault scarp with the rotation occurring away from the fault plane. These features were few in number and occur primarily near the base of the east and west walls.

158 An artist concept of a series of tilted fault blocks (Figure 125i). This tectonic element occurs most frequently at the base of the east and west walls as well as on the walls themselves.

Fault Scarps of Inner Floor

Slump Notches

Not all of the tectonic features observed in the inner rift valley occurred within the bedrock terrain. When traversing the talus accumulations, commonly present at the base of both flow units and fault scarps, *slump notches* (Figure 159) were seen. The exact origin of the features was not clear as they could have resulted from movement along buried fault planes or from regional slumping and mass wasting processes.

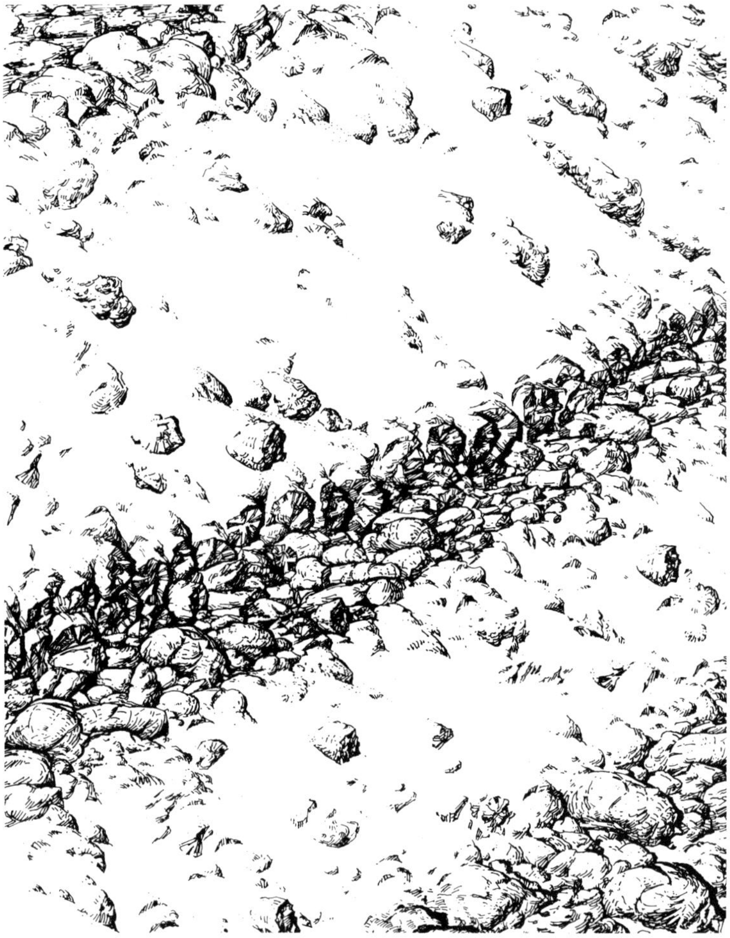

159 An artist concept of a slump notch (Figure 125h). These features are difficult to recognize in the field and were most frequently seen in talus accumulations where the nature of bedrock movement is hard to detect.

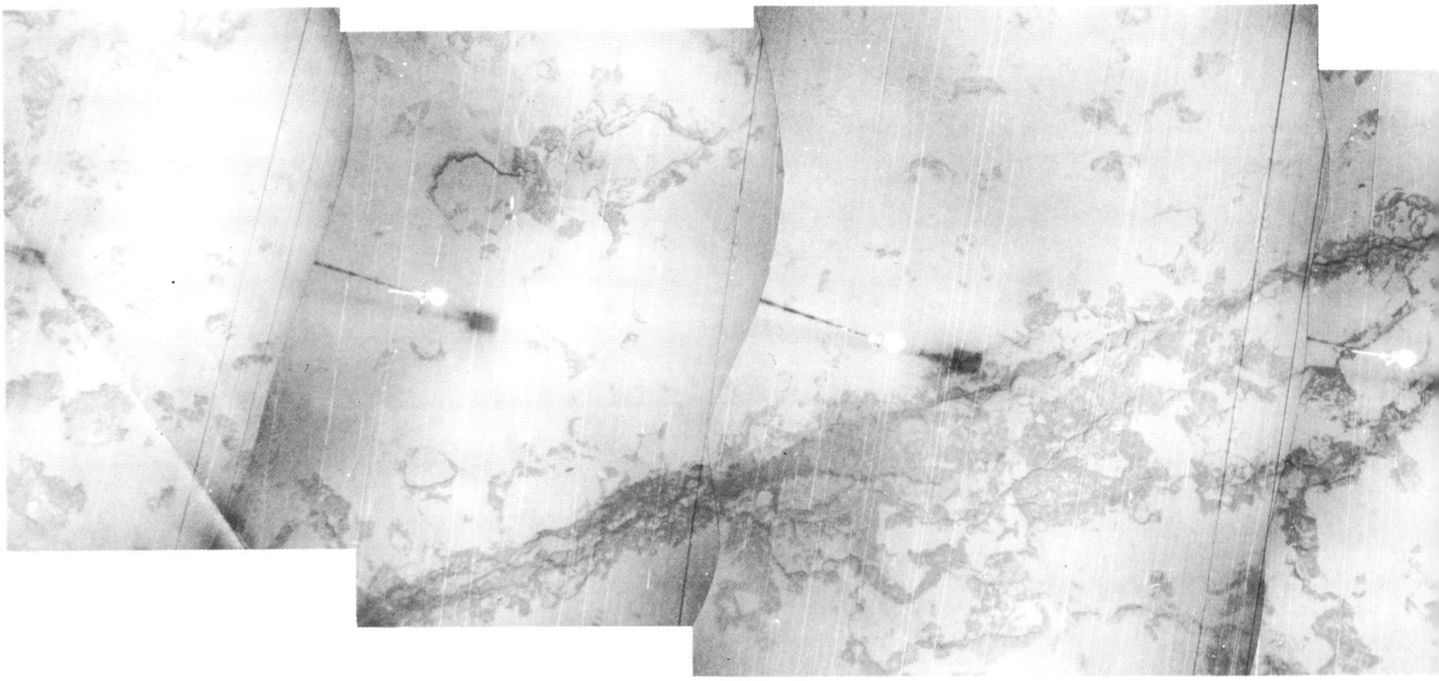

Patterns of Faulting

Although faults and fissures occur throughout the inner rift valley at an average spacing of 30 meters, certain areas—particularly at the base of volcanic flows and near their crests—are characterized by a closely spaced network of faults commonly producing a highly brecciated series of fault blocks. The dominant fault planes in these areas parallel the rift axis, but many of the faults are curved, branch, or consist of short *en echelon* segments. The LIBEC pictures when grouped together into mosaics provide some of the best pictures of these fault patterns (Figures 160 and 161). It was possible in some cases, however, to construct similar mosaics of these fault networks using pictures taken with the external camera on ALVIN (Figure 162).

160 LIBEC mosaic of two parallel fissures.

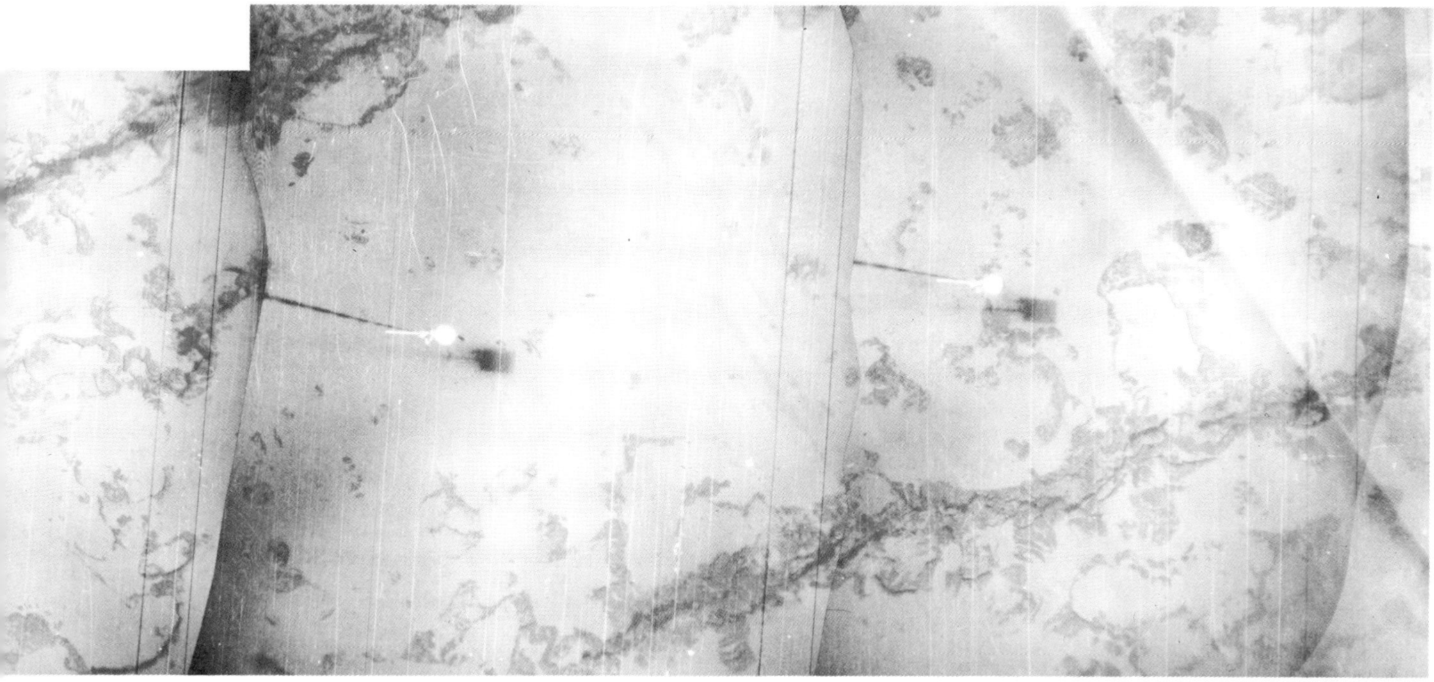

161 LIBEC mosaic showing a swarm of fissures.

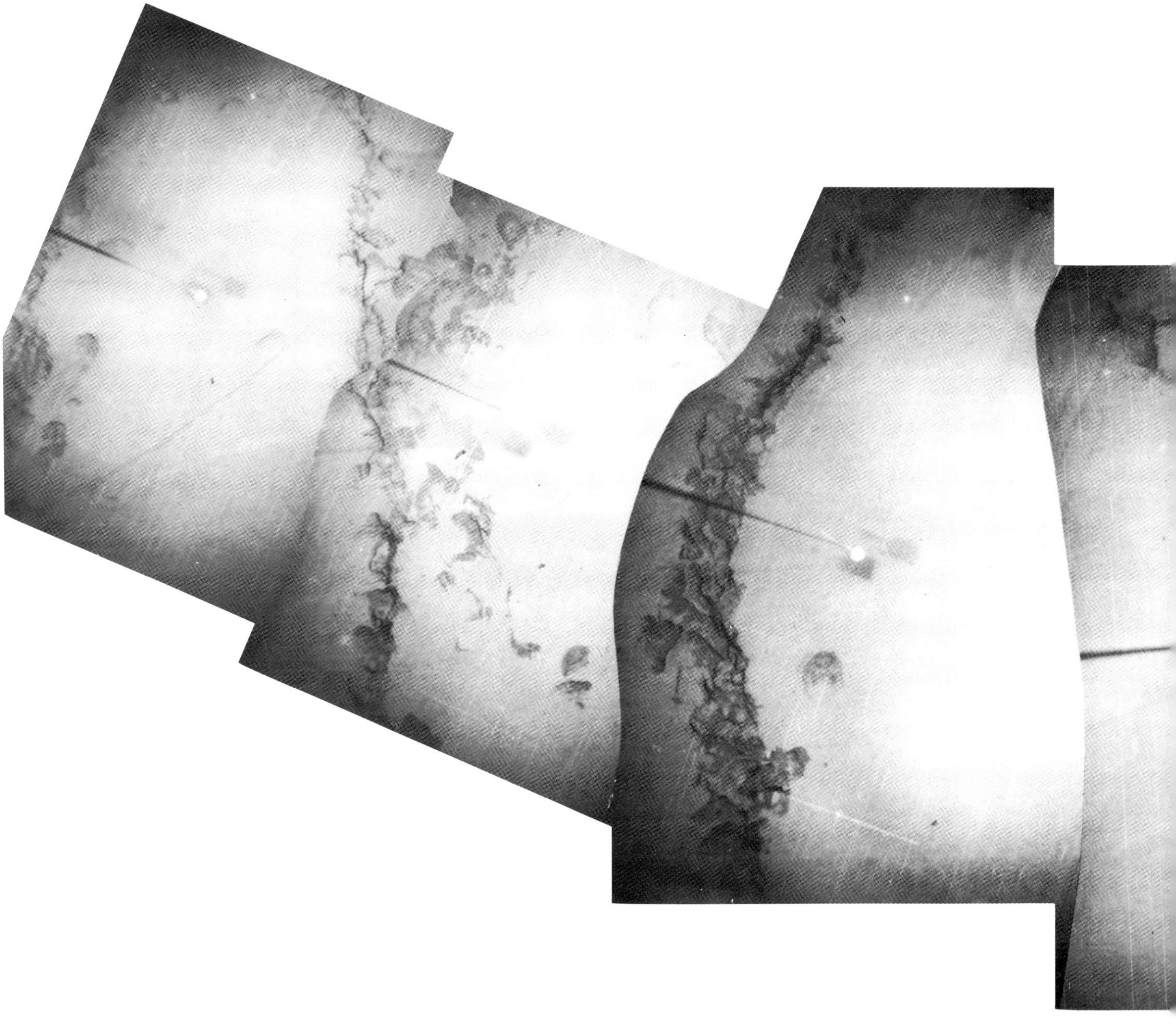

162 Mosaic of photographs taken from ALVIN. The fissure in the center is approximately 4 meters wide. The insert shows the fracture pattern observed in the mosaic.

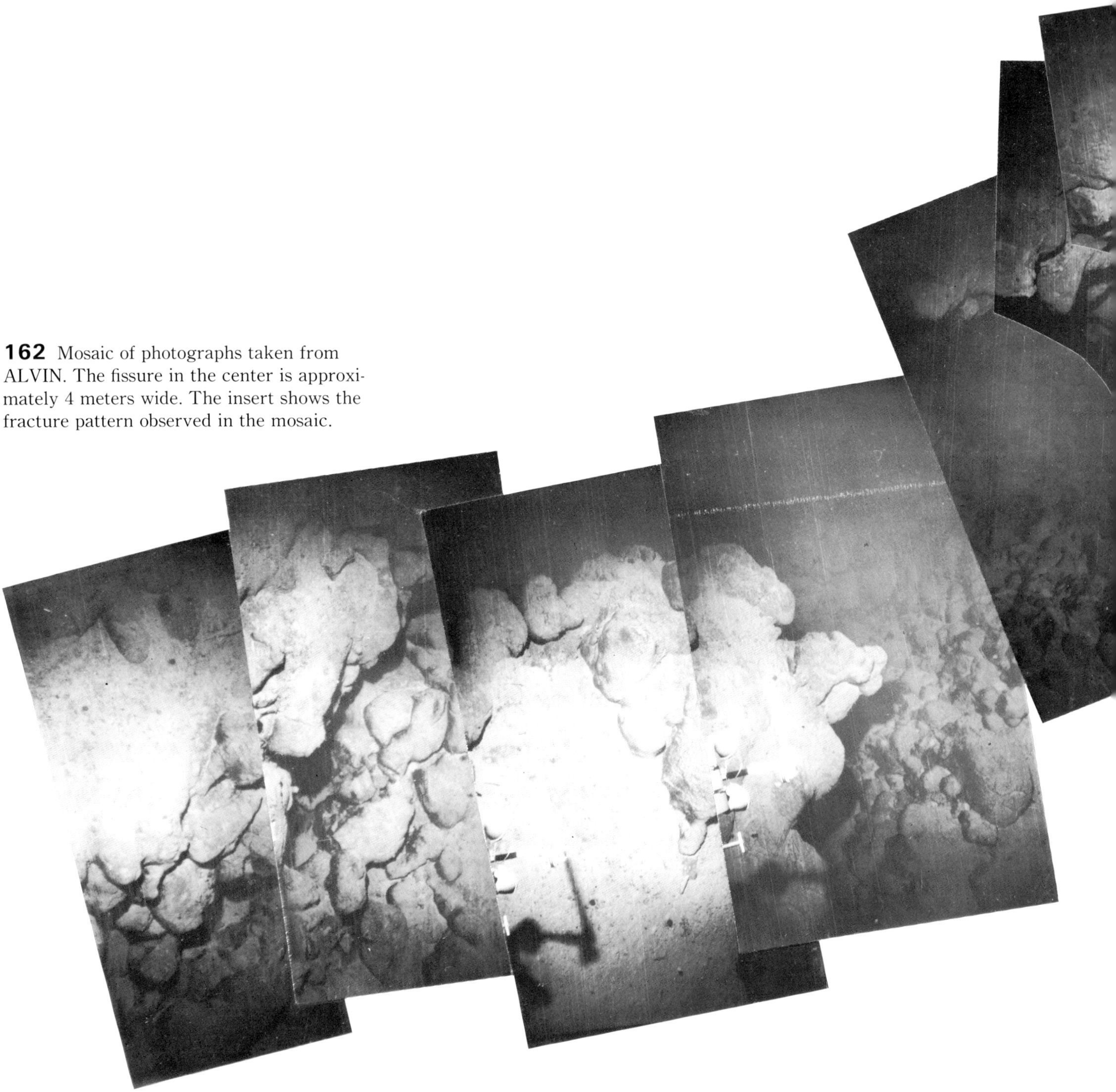

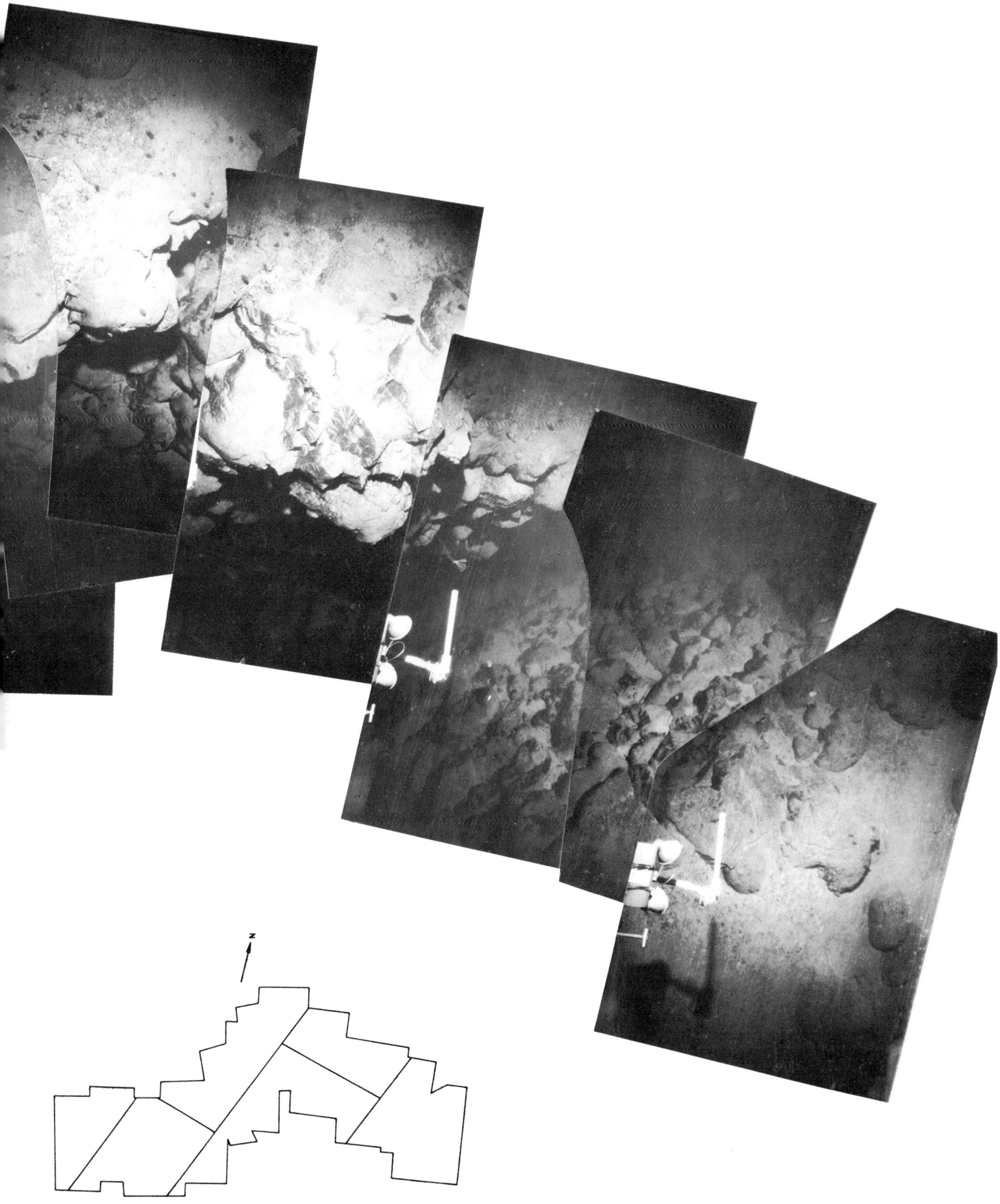
N

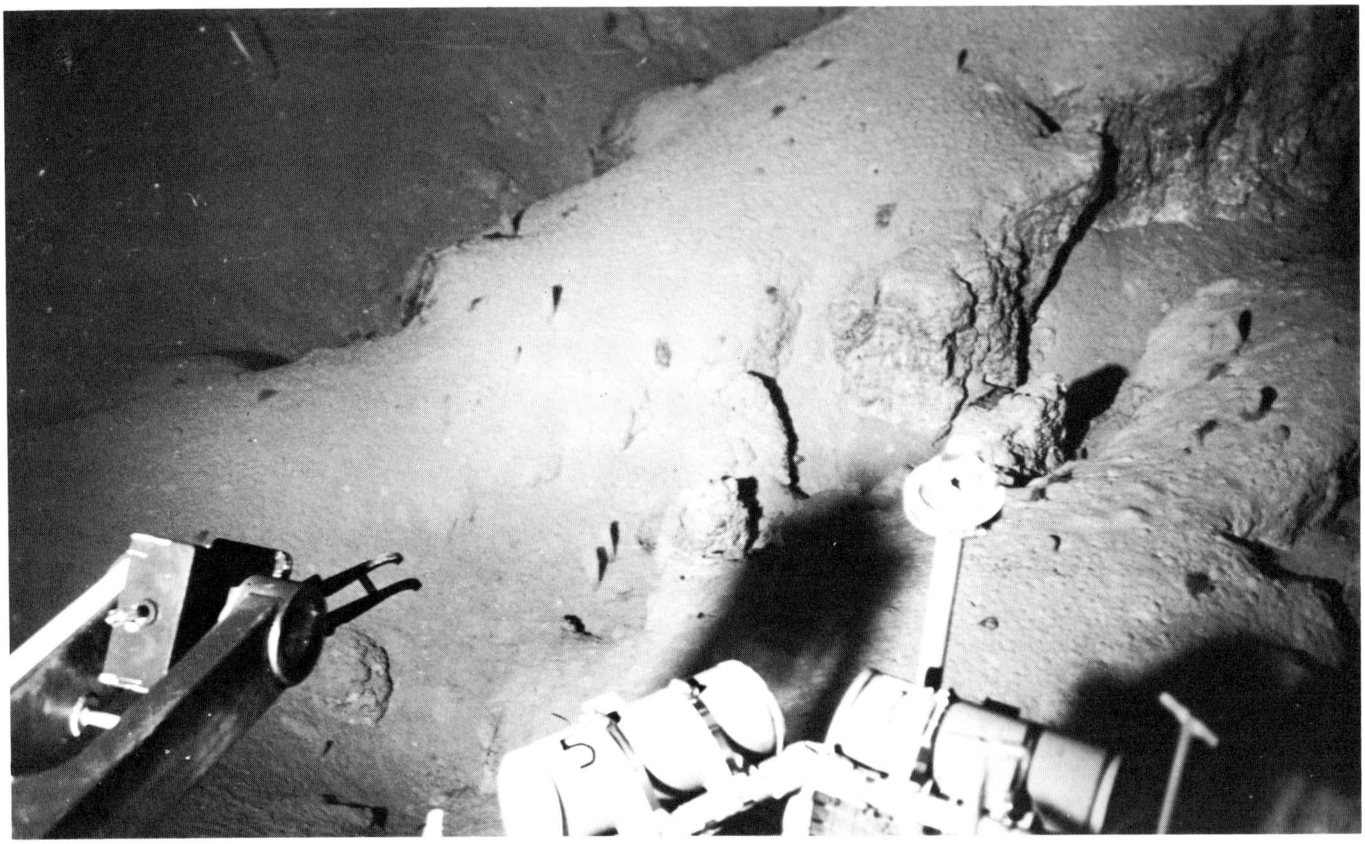

163 Faulted lithified sediments encountered at the base of the west wall.

Faulted Sedimentary Rock

Despite the presence of sedimentary accumulations in the inner rift valley floor and the apparent periodic nature of volcanic activity, no examples of pillow lavas overlying sedimentary layers were found although a few examples have been reported by the French near the base of Mt. Venus (Chemineé personal communication). In some places, however, faulted lithified sediments were observed (Figure 163). These occur at the base of valley walls where sediment accumulations are frequently greater than several meters thick and vertical faulting is prevalent as well as in some of the sediment-filled depressions. (Figures 164 and 165).

164 Outcrop of sedimentary material at the base of the west wall. Material was of sufficient hardness that it could be sampled using the plastic core being held by the mechanical arm in this photograph.

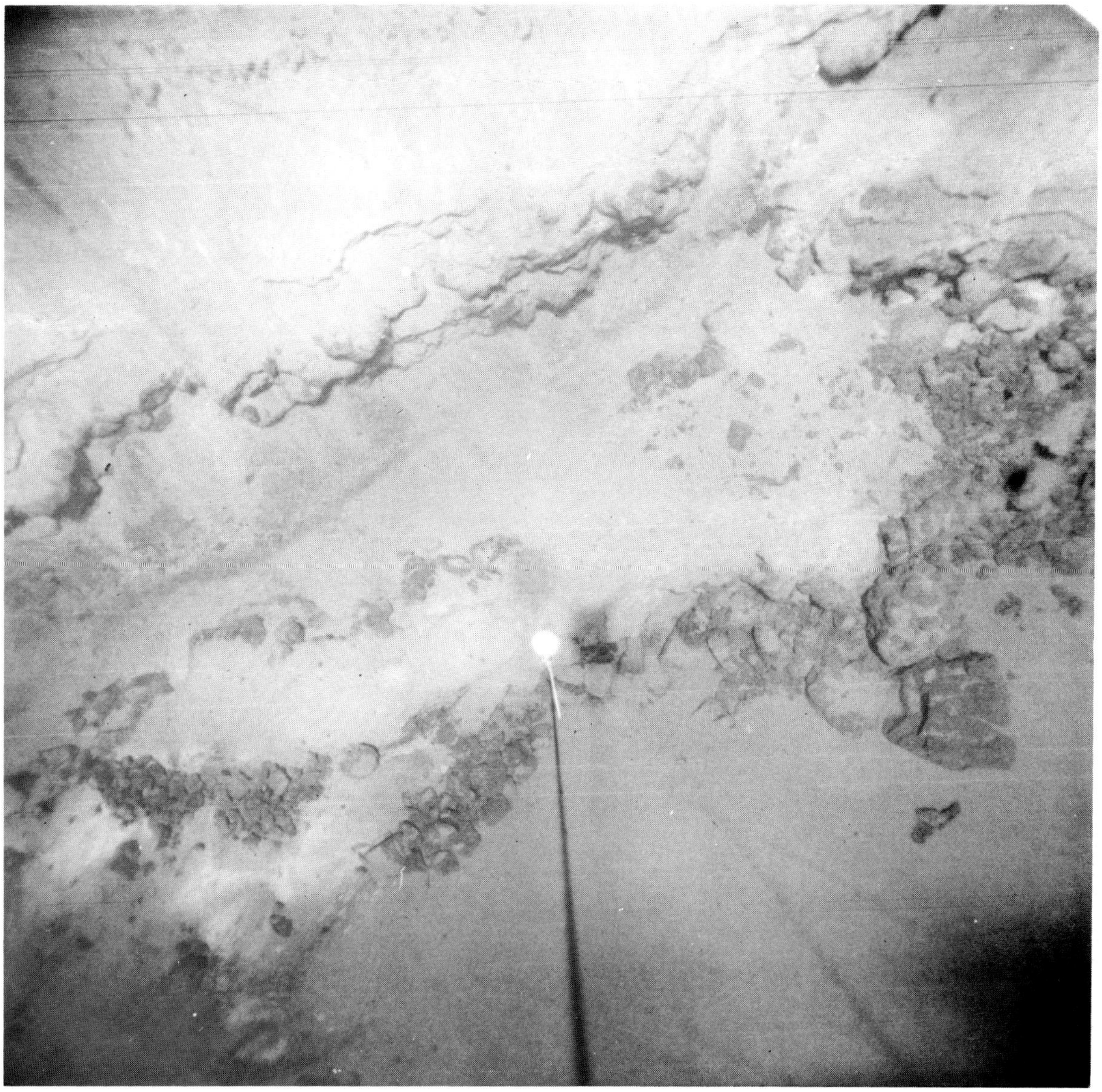

165 An outcrop of sedimentary material similar to that shown in Figure 164 except in this case the view is from directly overhead using the LIBEC camera.

Rift Valley Walls

The nature of faulting on the east and west valley walls contrasts markedly to that of the inner floor. The open tensional fissures with small-scale vertical offset are absent. In their place is a series of steep fault scarps with dips averaging 60° and vertical relief exceeding 300 meters in places. At the base of the west wall is a steep sediment-covered ramp (Figure 166). The sediment is several meters thick and a wide variety of animal burrows and tracks are present (Figure 167). In some areas, particularly in small depressions in the sediment ramp, large accumulations of pteropod shells are found (Figure 168). The shells are hollow, covered with a manganese stain, and are nearly neutrally buoyant. Only a small amount of turbulence (from the submersible propeller) was necessary to shape the pteropod accumulations into large ripple marks.

Upward on this sediment ramp towards the wall, progressively more talus fragments are mixed with the sediment until talus covers the entire slope (Figure 169). The dominant fragments are small fist-sized blocks exhibiting a radial joint pattern characteristic of fractured pillow lavas. Unlike the talus accumulations of the inner floor, however, these pillow fragments are intermixed with large blocks of massive basalt up to 1–2 meters in size.

The top of the fault scarps on the west wall begin with a back tilted surface (Figure 170). The surface is generally covered with talus blocks mixed with sediments (Figure 171) while in other places the surface is of volcanic origin and the edge of the scarp is a sharp boundary (Figure 172). A short distance down the scarp face, truncated pillows having radial cross sections are encoun-

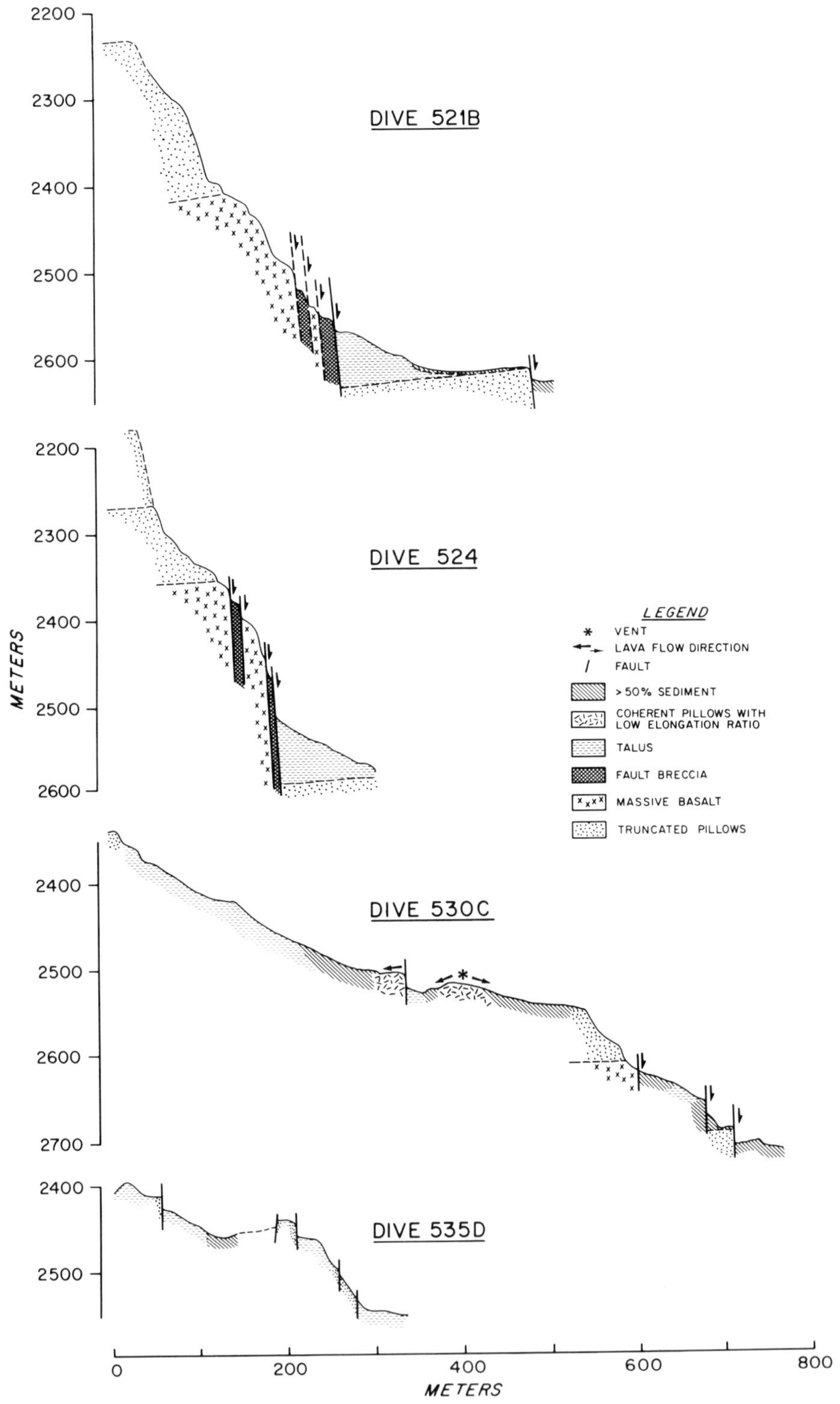

166 Geologic traverses of four dives conducted by ALVIN up the face of the west wall. (After Ballard and van Andel, 1977.)

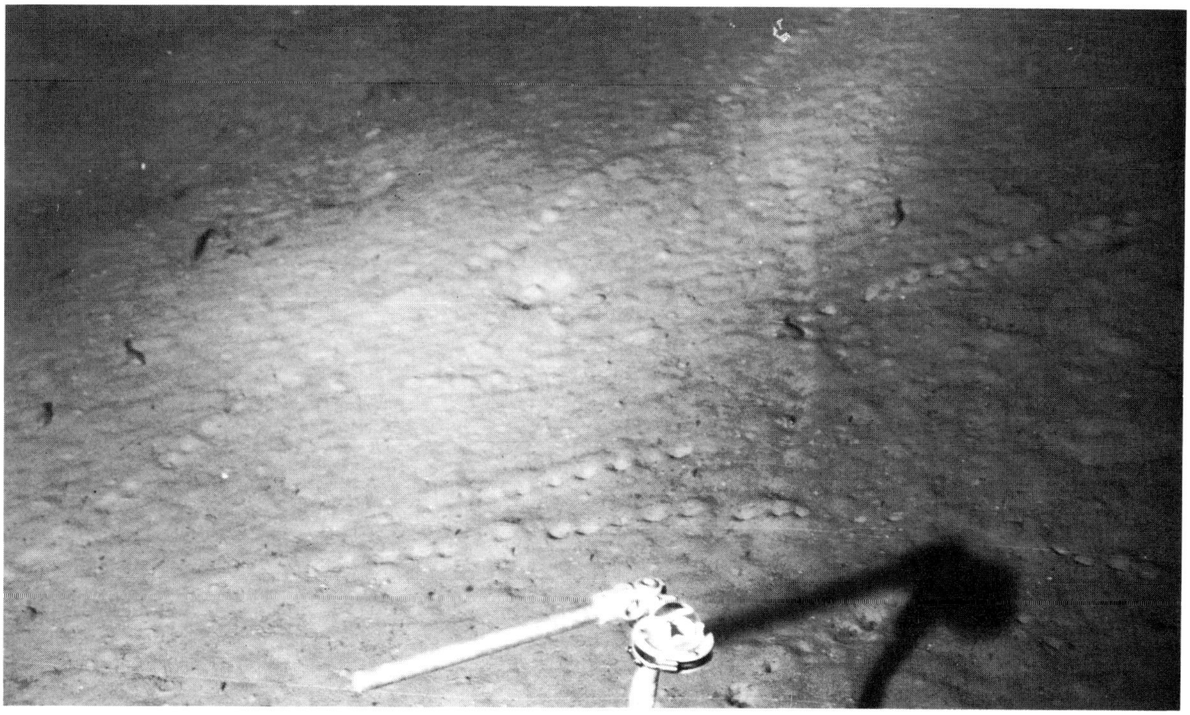

167 Sediment accumulations reached their maximum coverage and thickness at the base of the west wall where large areas are totally covered with sediments. The small depressions shown in this photograph are the result of burrowing activity by benthic organisms.

168 Thick accumulations of hollow pteropod shells commonly occur on the sediment surface. In this photograph they appear as dark patches while in other regions they have been shaped into large ripple marks. Their dark appearance is the result of a manganese coating on the hollow shells.

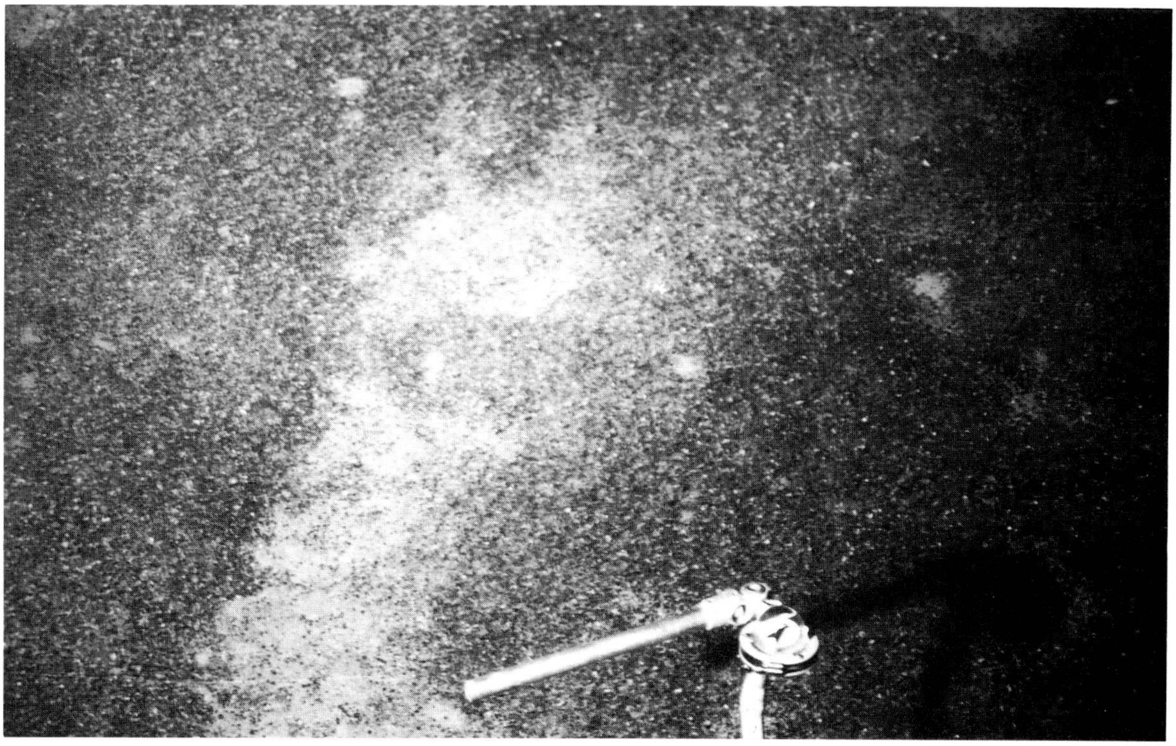

169 A major talus ramp occurs at the base of the west wall. The great size and extent of these accumulations are in marked contrast to the fault scarps of the inner floor and suggest more violent activity associated with vertical uplift in this tectonic province.

170 A back tilted surface at the top of a fault scarp on the west wall. The amount of tilting averaged between 5° and 7°.

171 Top of a tilted fault scarp on the west wall covered with talus blocks and sediments.

172 LIBEC photograph looking directly down on a small bench or terrace on the southeast slope of Mt. Venus similar to those seen on the west wall. A secondary fracture can be seen near the edge of the scarp.

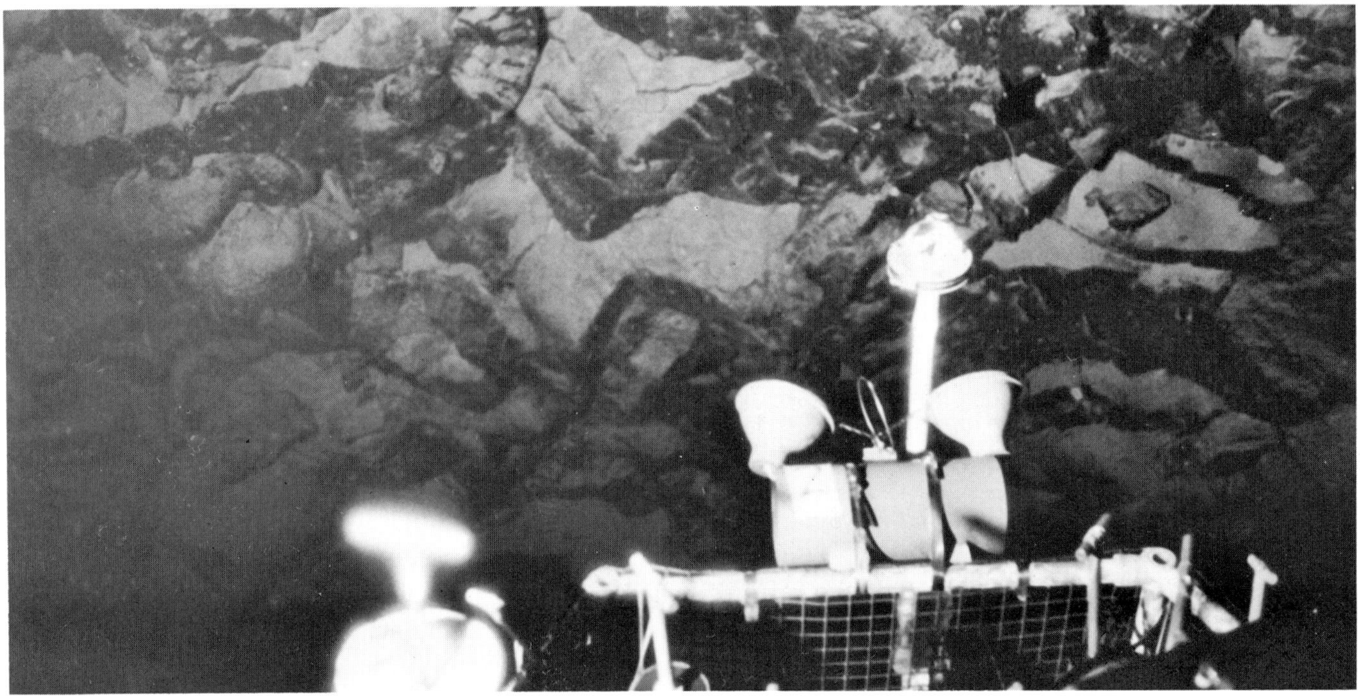

173 Truncated pillows near the top of the section exposed in the west wall (Figure 166). The section is identical to that observed in the smaller fault scrap in the valley below.

tered (Figure 173). The nature of the truncated pillow surface is identical to those observed in the valley floor and the ratio of truncated to coherent pillows varies widely (Figure 174). At approximately 180 meters from the top of the fault scarp, the truncated pillows begin intermixing with a more massive basalt section (Figure 175). About 20 meters lower this transition is complete. Near this transition zone, as well as at greater depths in the section, a series of lenticular intrusive bodies are found (Figure 176). Their width varies between 5–30 cm, and their strike is at times vertical while in other instances it is close to horizontal. The outcrop surface along the massive basalt section is complex containing slump scarps, fault reentrants (Figure 177), vertical striations, slickenslides, and fault gouge. The fault plane is nearly vertical in places and straight, while in other regions it is curved (Figure 178).

174 Mixture of coherent and truncated pillow forms occur in the west wall pillow section.

175 Pillow lava features begin to blend with a more massive basalt texture at approximately 180 meters into the section (Figure 166).

176 A lenticular intrusive cutting horizontally across the massive basalt. The intrusives have a closely jointed structure suggesting a more rapid cooling history than the rock they intrude.

The massive basalt section traversed, averaged 125 meters in thickness. Intermixed with the massive basalt and extending deeper into the section are exposures of what appeared to be welded breccia (Figure 179). The welded breccia occurs in sheets and exhibits subtle horizontal bedding planes as well as distinctive contacts between layers of varying courseness (Figure 180). The breccia has a fine-grained matrix of light colored material surrounding larger nodular-shaped rock fragments and is striated by a series of vertical fault gouges (Figure 181). The origin of the breccia is unclear, but several field relations suggest it may be related to fissure formation in the valley floor. The small horizontal thickness of the breccia sheets, the fault planes bounding them, and the composition of the breccia itself suggests that the numerous open fissures (Figure 182) observed in the inner floor are in time partially or completely filled with talus and sediment (Figure 183). When a portion of the newly formed crust undergoes vertical uplift on the inward facing valley walls, these fissures are apparently reactivated and act as the fault planes along which major dip-slip displacement occurs. During this vertical uplift, the sediment and talus are ground together and compressed to form the coherent breccia sheets. The French scientists traversed the west wall to the north of where our observations were made and did not encounter breccia sheets. They did, however, observe sediment layers interbedded and baked by overlying pillow flows (Chemineé personal communications).

177 The fault surface contains numerous fault reentries, particularly within the massive basalt section.

178 A photograph taken with the LIBEC camera looking directly down the face of a fault scarp, within the inner floor revealing its rough and curved surface. Similar fault planes were seen on the west wall.

179 Welded breccia at the base of the west wall. The matrix is a mixture of fine-grain dark and light material surrounding rounded rock fragments.

180 A contact within the welded breccia section between layers of different coarseness. If the breccia is derived from filled fissures, these contacts may represent sudden changes in the fissure's depositional history.

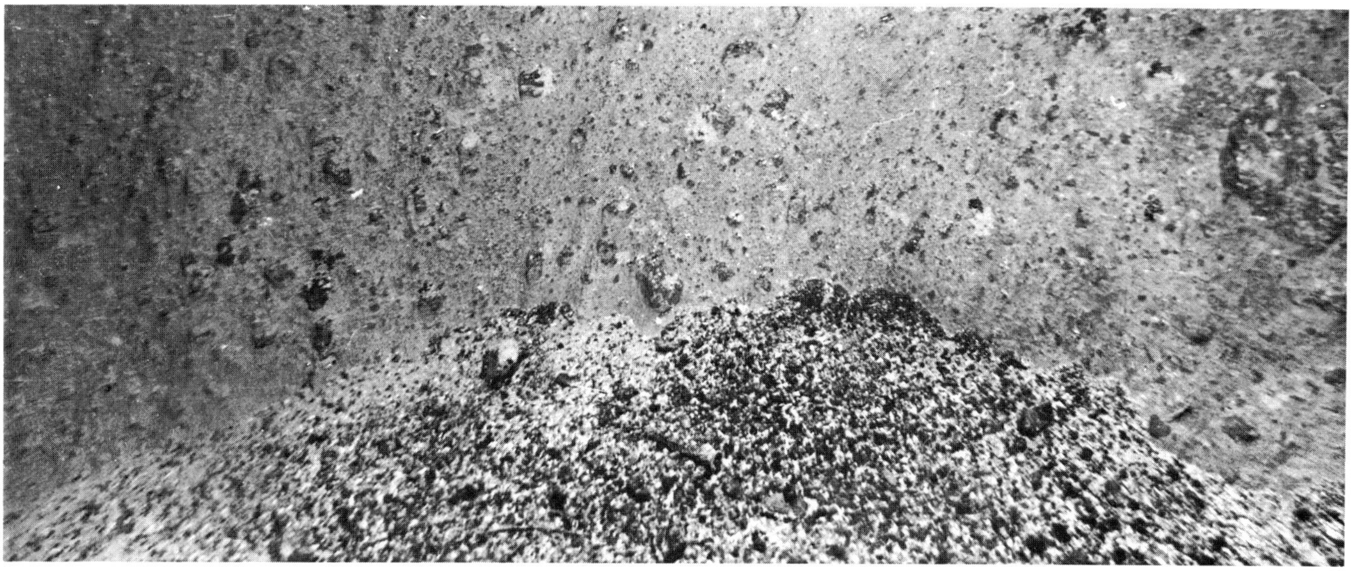

181 Fault gouge is most prevalent at the base of the west wall within the welded breccia section.

182 An open fissure in the rift valley floor actively being filled with sediments.

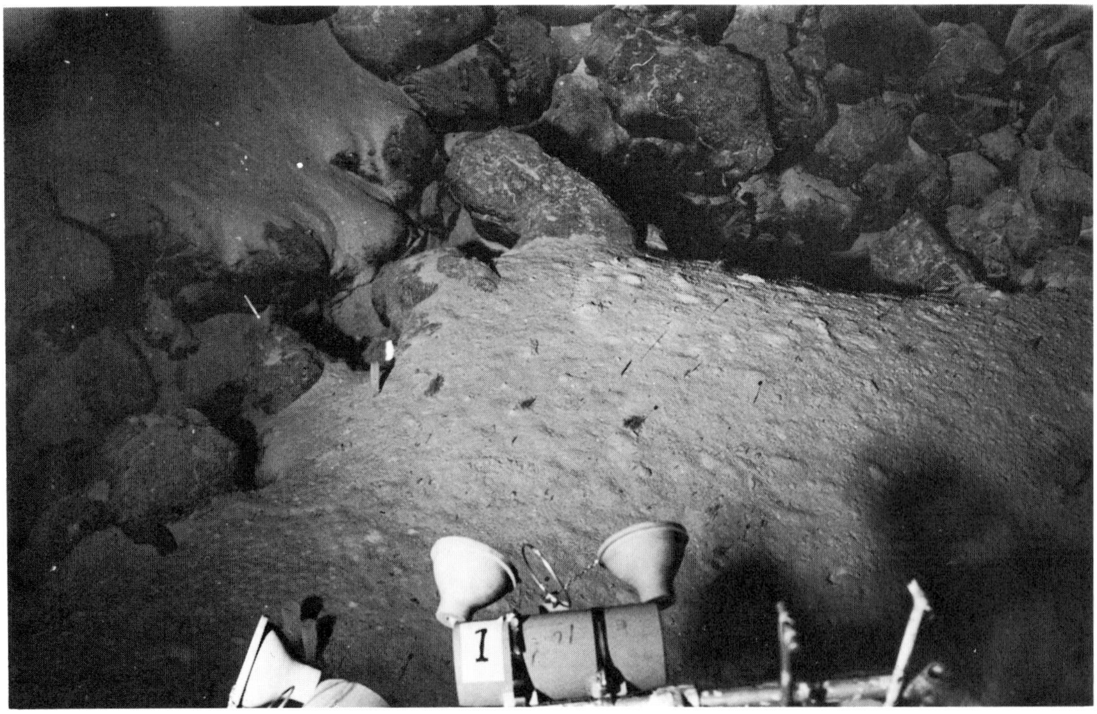

183 A fissure which has been completely filled with sediment and talus fragments. If this fault plane is reactivated during vertical uplift on the valley walls, the fissure filling may be compressed and ground to form the welded breccia observed at the base of the wall.

References Cited

Arcyana, X. Y. 1975. Direct sea-floor observations in submersibles of an active rift valley and transform fault in the Atlantic near 36°50′ N. *Science* 190: 108–116.

Ballard, R. D., and van Andel, Tj. 1977. Project FAMOUS: Morphology and tectonics of the inner rift valley at 36°50′ N on the Mid-Atlantic Ridge. *Geol. Soc. Amer.* Bull. 88: 507–530.

Ballard, R. D. and van Andel, Tj. 1977. Project FAMOUS: Operational techniques and American submersible operations. *Geol. Soc. Amer. Bull.* 88: 495–506.

Ballard, R. D., Bryan, W. B., Heirtzler, J. R., Keller, G., Moore, J. G., and van Andel, Tj. H. 1975. Manned submersible observations in the FAMOUS area, Mid-Atlantic Ridge. *Science* 190: 103–108.

Bellaiche, G. and Cheminée, J. L. 1976. Les formes observées pendant l'opération "FAMOUS" dans le plancher du rift Médio-Atlantic par 36°50′N, C. R. Acad. Sci. Paris. 282: 519–522.

Bellaiche, G., and Cheminée, J. L., Francheteau, J., Hekinian, R., Le Pichon, X., Needham, H. D., Ballard, R. D. 1974. Inner floor of the rift valley: first submersible study. *Nature* 250: 558–560.

Brundage, W. L., Jr., and Cherkis, N. Z. 1975. Preliminary LIBEC/FAMOUS cruise results. Naval Research Laboratory Report 7785 (unpublished manuscript) 31 pp.

Bullard, E. C., Everett, J. E., and Smith, A. G. 1965. A symposium on continental drift IV: The fit of the continents around the Atlantic. *Royal Soc. London Phil. Trans.* 258: 41–51.

Dietrich, V. 1967. Geosynklinaler Vulkanismus in den oberen penninischen Decken Graubundens (Schweiz). *Geol. Rundsch.* 57: 246–264.

Duffield, W. A. 1969. Concentric structure in elongate pillows, Amador County, Calif.: *U.S. Geol. Survey Prof. Paper* 650-D, pp. D19–D25.

Fuller, R. E. 1931. The aqueous chilling of basaltic lava on the Columbia plateau. *Am. Jour. Sci.,* ser. 5, vol. 21, no. 125, pp. 281–300.

Heirtzler, J. R., and Bryan, W. B. 1975. The floor of the Mid-Atlantic rift. *Scientific American,* vol. 233, no. 2, pp. 78–90.

Heirtzler, J. R. and Le Pichon, X. 1974. FAMOUS: A plate tectonics study of the genesis of the lithosphere. Geology, vol. 2, pp. 273–274.

Hekinian, R., Chaigneau, M., and Cheminee, J. L. 1973. Popping rocks and lava tubes from the Mid-Atlantic rift valley at 36°N. *Nature* 245: 371–373.

Larson, R. L. 1971. Near bottom geologic studies of the East Pacific Rise crest. *Geol. Soc. Am. Bull.* 82: 823–842.

Luyendyk, B. P. and Macdonald, K. C. 1977. Physiography and structure of the FAMOUS rift valley inner floor observed with a deeply-towed instrument package. *Geol. Soc. Amer. Bull.* 88: 648–663.

Macdonald, K. C. 1977. Near-bottom magnetic anomalies, asymmetric spreading, oblique spreading and tectonics of the accreting plate boundary on the Mid-Atlantic Ridge near 37°N (FAMOUS). *Geol. Soc. Amer. Bull.* 88: 541–555.

Moore, J. G. 1975. Mechanism of formation of pillow lava. *Am. Scientist,* vol. 63 no. 3, pp. 269–277.

Moore, J. G., Cristofolini, R., and Lo Giudice, A. 1971. Development of pillows on the submarine extension of recent lava flows Mt. Etna, Sicily. *U.S. Geol. Survey Prof. Paper* 750-C, C80–C97.

Moore, J. G., Phillips R. L., Grigg, R. W., Peterson, D. W., and Swanson, D. A. 1973. Flow of lava into the sea 1969–1971, Kilauea Volcano, Hawaii. *Geol. Soc. Am Bull.* 84: 537–546.

Normark, W. R. 1976. Delineation of the main extrusion zone of the East Pacific Rise at 21°N latitude. *Geology,* 4: 681–685.

Phillips, J. D., and Forsyth, D. 1972. Plate tectonics, paleomagnetism, and the opening of the Atlantic. *Geol. Soc. Amer. Bull.* 83: 1579–1600.

Ramberg, I. B., and van Andel, Tj. H. 1977. Morphology and tectonic evolution of the southern rift valley at 36°30′N in the FAMOUS area, Mid-Atlantic Ridge. *Geol. Soc. Amer. Bull.* 88: 577–586.

Ramberg, I. B., Gray, D. F., and Raynold, R. G. H. 1977. Tectonic evolution of a segment of the Mid-Atlantic Ridge: The FAMOUS area between 36°50′N and 37°20′N. *Geol. Soc. Amer. Bull.* 88: 609–620.

Tepley, L. (producer) and Moore, J. G. (sci. consultant) 1974. Fire Under the Sea: The Origin of Pillow Lava (16 mm sound motion picture). Moonlight Productions, Mountain View, Calif.

Waters, A. C. 1960. Determining direction of flow in basalt. *Am. Jour. Sci.,* v. 258–A, pp,. 350–366.